Chinese and Russian Perceptions of and Responses to U.S. Military Activities in the Space Domain

ALEXIS A. BLANC, NATHAN BEAUCHAMP-MUSTAFAGA,
KHRYSTYNA HOLYNSKA, M. SCOTT BOND, STEPHEN J. FLANAGAN

RAND NATIONAL SECURITY RESEARCH DIVISION

For more information on this publication, visit **www.rand.org/t/RRA1835-1**.

Published by the RAND Corporation, Santa Monica, Calif.

Library of Congress Cataloging-in-Publication Data is available for this publication.

ISBN: 978-1-9774-1026-9

Cover: composite design by Carol Ponce; adapted from images by Kjell Lindgren/NASA, Nucia/AdobeStock.

About This Report

Since the 1980s, U.S. military activities and policy with respect to the space domain have evolved significantly. The United States established U.S. Space Command in 1985 as a functional combatant command to integrate the various U.S. military services' space-related activities and implement the Reagan administration's Strategic Defense Initiative. This command was, however, merged with U.S. Strategic Command in 2002 during the focus on homeland defense after the terrorist attacks on September 11, 2001. More-recent developments include the reestablishment of U.S. Space Command and the establishment of the U.S. Space Force in 2019, and multiple high-profile policy and warfighting documents have followed. Given this activity and the concerns raised in a number of domestic and international fora regarding the increasingly congested and contested nature of space, there has been surprisingly little open-source analysis of Chinese and Russian perceptions of these developments.

This report fills that gap by exploring native-language primary sources, such as Chinese and Russian government publications, military journals, academic reports, and domestic media, that convey internal perceptions of developments in U.S. military space and counterspace doctrine, exercises, and military organization from 1983 (when the Strategic Defense Initiative was proposed) to 2014 (when the Geosynchronous Space Situational Awareness Program was revealed) to understand how Chinese and Russian perceptions about U.S. military activities in the space domain have evolved over time and what responses China and Russia appear to be taking to address or counter U.S. actions.

RAND National Security Research Division

This research was sponsored by the U.S. government and conducted within the International Security and Defense Policy Program of the RAND National Security Research Division (NSRD). NSRD conducts research and analysis for the Office of the Secretary of Defense, the U.S. State Department, allied foreign governments, and foundations.

For more information on the RAND International Security and Defense Policy Program, see www.rand.org/nsrd/isdp or contact the director (contact information is provided on the webpage).

Acknowledgments

We greatly valued the support of several individuals in the International Security and Defense Policy Program of NSRD, including Michael McNerney, Michael Spirtas, Molly Dunigan, Mark Cozad, Jim Mitre, and Norma Ockershausen. We benefited from thoughtful reviews by Phillip Saunders and Frank Klotz.

Summary

Issue

Since the 1980s, U.S. military activities and policy with respect to the space domain have evolved significantly. The United States established U.S. Space Command in 1985 as a functional combatant command to integrate service space activities and implement the Strategic Defense Initiative (SDI). More-recent developments include the reestablishment of U.S. Space Command and the establishment of the U.S. Space Force in 2019, and multiple policy and warfighting documents have rapidly followed. Given this activity and the concerns raised in domestic and international fora regarding the increasingly congested and contested nature of space, there has been surprisingly little open-source analysis of Chinese and Russian perceptions of these developments.

Approach

To fill this gap, we focused on native-language primary sources, such as Chinese and Russian government publications, military journals, academic reports, and domestic media. We used these sources as a broadly representative sample of materials that convey internal perceptions of developments in U.S. space and counterspace doctrine, exercises, and military organization. Working with the sponsor, we developed a representative sample of ten U.S. "events" in the space domain to focus our efforts. We then reviewed the primary sources for discussions of these events to assess how Chinese and Russian reactions to U.S. military activities related to space have evolved over time.

Findings

The Chinese and Russian native-language primary sources reviewed for this report reflect a sustained perception that U.S. activities related to space are threatening and demonstrate hostile U.S. intent. This perception of hostile U.S. intent partly encompasses the space-based threat to their respective nuclear deterrents (e.g., from SDI), as well as concerns related to U.S. counterspace capabilities (e.g., Operation Burnt Frost) and concerns regarding the ability of U.S. satellites to fly closely (and covertly) to other space objects to inspect them and collect information (e.g., the Geosynchronous Space Situational Awareness Program).

The longer-term military *potential* of various U.S. activities regarding space, rather than the immediate impact of a specific activity or policy at that point in time, appears to be the key overarching factor shaping Chinese and Russian perceptions of the U.S. threat. Moreover, the condition of U.S.-Chinese or U.S.-Russian bilateral relations at a particular moment appears to shape each government's view of U.S. military space-related activities. Yet both countries tend toward confirmation bias, whereby more plausibly "aggressive" U.S. activities tend to reinforce the perception that the U.S. military has a "hostile" intent in the space domain, while more plausibly "cooperative" U.S. initiatives are discounted as disingenuous. China and Russia generally attempt to strike a nuanced rhetorical balance of characterizing U.S. actions as threatening while characterizing their own, similar actions as nonthreatening. China and Russia also appear to be concerned that technology might eventually catch up with perceived U.S. intentions to militarize and dominate space.

While it is difficult to definitively explain all the drivers of any country's space activities, it appears that Washington, Beijing, and Moscow are caught in an action-reaction cycle that perpetuates justifications for continued military actions in space based on previous adversary activities.

The following findings from our research are particularly noteworthy:

- One of the few instances in which a causal relationship can be drawn between a U.S. action and a Russian counteraction is the U.S. withdrawal from the Anti-Ballistic Missile (ABM) Treaty in 2002. It appears that Russian leaders pushed scarce resources back into their ballistic missile defense programs; developed requirements for many new types of weapons, including anti-satellite weapons and munitions; and developed a unified aerospace defense system.
- Both Chinese and Russian officials have contended that the United States has led the way to militarizing space, particularly since 2001, leaving them no choice but to take countervailing measures. Both countries point to the U.S. decision to withdraw from the ABM Treaty as a key inflection point in U.S. efforts to weaponize space.
- China and Russia are highly sensitive to the perception that arms racing is a negative behavior, and both emphasize the need to avoid such behavior because it makes the world less safe and carries immense costs.

Finally, taking a step back, it is striking that, in selecting a representative set of pacing events—actions taken by the United States that might plausibly have influenced Chinese and Russian perspectives of the space domain—we broadly "got it right" with respect to Russia. However, some events that mattered to Russia were essentially nonevents in Beijing, and three U.S. actions that were not part of the sample received a great deal of attention in China. This result suggests that the diplomatic history and cultural understanding between the United States and the Union of Soviet Socialist Republics/Russia was perhaps more formative than might be fully appreciated. This history appears to allow the United States to manage its relationship more effectively with Russia—or at least to understand in retrospect which U.S. actions get Moscow's attention. It is a separate question, outside the scope of this study, whether the United States accurately predicts how its policies and actions might be perceived by Russia. Such a history and cultural understanding is nearly absent in the U.S. relationship with China. Our analysis suggests that U.S. policymakers should be relatively modest regarding the U.S. ability to anticipate or manage Chinese perceptions of U.S. actions in the space domain.

Contents

Figures

Chapter 1. Exploring Perceptions About U.S. Military Activities in the Space Domain

The Issue

The United States has undertaken a variety of initiatives in the space domain since 2015—backed by a significant budgetary allocation and multiple high-visibility actions since 2019. For example, the Trump administration issued the "America First" U.S. *National Space Strategy* in 2018 and the National Space Policy in 2020. In addition, in 2019, U.S. Space Command was reestablished and the U.S. Space Force (USSF) was established. A multitude of high-profile policy and warfighting documents followed. In 2020, the U.S. Department of Defense (DoD) simultaneously issued its *Space Strategy* and introduced U.S. Space Command's Space Capstone Publication, *Spacepower*, and the Joint Staff released an updated doctrine for *Space Operations*, Joint Publication 3-14. Statements by U.S. officials in domestic and international fora have also pounded a steady drumbeat, emphasizing their concerns regarding the congested, contested, and competitive nature of space.[1]

For their part, both Beijing and Moscow have denounced in numerous fora what they characterize as U.S. efforts to "weaponize," "militarize," and "dominate" space. In its 2019 defense white paper, China criticized the United States, saying that it has "provoked and intensified competition among major countries, . . . pushed for additional capacity in nuclear, outer space, cyber and missile defense, and undermined global strategic stability."[2] A 2021 Chinese submission to the United Nations on space norms similarly condemns Washington:

> [T]he weaponization of and an arms race in outer space becomes more prominent and pressing. The root cause is that [a] certain country [the United States] sticks to the Cold War mentality, pursues unilateral military and strategic superiority in space, and increase[s] its attempts, plans and actions to seek dominance in space.[3]

Among the major threats to the Russian Federation listed in Russia's most recent *National Security Strategy*, the threat of space being "actively explored" as a "new sphere of warfare" received attention, as did the need to "ensure the interests of the Russian Federation related to the

---

[1] U.S. Senate, "Testimony on Space Force, Military Space Operations, Policy and Programs," Washington, D.C., U.S. Senate Committee on Armed Services, May 26, 2021.

[2] People's Republic of China (PRC) Information Office of the State Council, "China's National Defense in the New Era," Xinhua, July 24, 2019.

[3] Permanent Mission of the People's Republic of China to the United Nations, "Document of the People's Republic of China Pursuant to UNGA Resolution 75/36 (2020)," April 30, 2021.

development of outer space . . . ."[4] The Kremlin has also ramped up its public denunciations of the militarization of space in recent years, emphasizing the global danger associated with such a "violation of international law."[5]

Despite resurgent activity by the United States, and despite increasingly confrontational rhetoric and actions by China and Russia, there has been a dearth of analysis of native-language Chinese or Russian primary sources tracing the evolution of either government's perceptions regarding U.S. military activities in the space domain. By and large, existing analysis has focused on the evolution of externally focused documents released by Beijing and Moscow—i.e., English-language writings regarding U.S. military actions. Existing scholarship has also focused on the evolution of Chinese or Russian space activities themselves. This ground has been well trod; therefore, these aspects of the subject are not central elements of this report.[6] Instead, this report adds a new facet to the existing research, examining native-language articulations of Chinese and Russian perceptions of U.S. military activities in the space domain over time.

Finally, this report does not address in depth how China and Russia discuss U.S. space activities in external messaging for perception-shaping. Multiple analyses have demonstrated that China and Russia have "gone to school" on the U.S. way of war in the wake of U.S. and

---

[4] Security Council of the Russian Federation, *Strategiya natsionalnoi bezopasnosti Rossiiskoi Federatsii*, [*The National Security Strategy of the Russian Federation*], Moscow, 2021.

[5] Kremlin, "Interv'yu teleradiokompanii 'Mir,'" ["Interview with the Television and Radio Company 'Mir'"], April 12, 2017; Kremlin, "Vstrecha s pomoshchnikom Prezidenta SSHA po natsbezopasnosti Dzhonom Boltonom," ["Meeting with Assistant to the President of the United States for National Security Affairs John Bolton"], October 23, 2018; and President of the Russian Federation, *Voyennaya doktrina Rossiyskoy Federatsii* [*Military Doctrine of the Russian Federation*], December 25, 2014.

[6] For recent studies on China's space activities, see Kevin Pollpeter, "The Chinese Vision of Space Military Operations," in James Mulvenon and David Finkelstein, eds., *China's Revolution in Doctrinal Affairs: Emerging Trends in the Operational Art of the Chinese People's Liberation Army*, Alexandria, Va.: CNA Corporation, 2005; Mark A. Stokes and Dean Cheng, *China's Evolving Space Capabilities: Implications for U.S. Interests*, Washington, D.C.: U.S.-China Economic and Security Review Commission, April 26, 2012; Kevin Pollpeter, Eric Anderson, Jordan Wilson, and Fan Yang, *China Dream, Space Dream: China's Progress in Space Technologies and Implications for the United States*, Washington, D.C.: U.S.-China Economic and Security Review Commission, March 2, 2015; Dean Cheng, Peter Garretson, Namrata Goswami, James Lewis, Bruce W. MacDonald, Kazuto Suzuki, Brian C. Weeden, and Nicholas Wright, *Outer Space; Earthly Escalation? Chinese Perspectives on Space Operations and Escalation*, ed. Nicholas Wright, Washington, D.C.: Joint Staff, U.S. Department of Defense, August 2018; Alexander Bowe, *China's Pursuit of Space Power Status and Implications for the United States*, Washington, D.C.: U.S.-China Economic and Security Review Commission, April 11, 2019; Brian Weeden, "Hearing on China in Space: A Strategic Competition?" testimony presented before the U.S.-China Economic and Security Review Commission, Washington, D.C., April 25, 2019; Mark Stokes, Gabriel Alvarado, Emily Weinstein, and Ian Easton, *China's Space and Counterspace Capabilities and Activities*, Washington, D.C.: U.S.-China Economic and Security Review Commission, March 30, 2020; and Brian Weeden, *Current and Future Trends in Chinese Counterspace Capabilities*, Paris: French Institute of International Relations, Proliferation Papers, No. 62, November 2020.

For recent U.S. government reports on Chinese space activities, see National Air and Space Intelligence Center, *Competing in Space*, Dayton, Ohio: Wright-Patterson Air Force Base, December 2018; Defense Intelligence Agency, *Challenges to Security in Space*, Arlington, Va., 2019, pp. 13–22; and Office of the Secretary of Defense, *Military and Security Developments Involving the People's Republic of China*, Washington, D.C.: U.S. Department of Defense, 2021.

allied interventions in Kosovo, Iraq, and Afghanistan, arguing that these events strongly influenced Chinese and Russian thinking about the future of warfare.[7] Nothing in our research would lead us to question those findings. Instead, we take a narrower approach: We assess internally focused Chinese and Russian analysis of U.S. military activities in the space domain—i.e., analysis written in the two countries' respective native languages. This analysis has the objective of answering, using publicly available information, the following questions:

- How have Chinese and Russian perceptions about U.S. military activities in the space domain evolved over time?
- What responses are China and Russia taking to address or counter U.S. actions?

Research Approach

To address the question of the evolution of Chinese and Russian perceptions of activities by the United States with respect to space, we focused on native-language primary sources, such as Chinese and Russian government publications, military journals, academic reports, and domestic media, that convey internal perceptions of developments in U.S. space and counterspace doctrine, policy, exercises, and military organization. Working with the sponsor, we developed a representative sample of U.S. "events" in the space domain to focus our efforts. This selection was intended to include both high-profile and lesser-known U.S. military space-related activities across a sufficiently lengthy period that publicly available, high-quality, Chinese and Russian writings addressing these activities would be available. Of note, the sample excludes U.S. civilian space activities, such as commercial use of space, space exploration, and human space flight, and other military applications, such as intelligence, surveillance, and reconnaissance; navigation and timing; and communications. We selected the following ten events:

- Strategic Defense Initiative (SDI) (1983) and U.S. Space Command creation (1985)
- President Bill Clinton's National Space Policy (1996)
- Mid-Infrared Advanced Chemical Laser (MIRACL) test (1997)
- Commission to Assess United States National Security Space Management and Organization ("Rumsfeld Commission") (2001)
- U.S. withdrawal from Anti-Ballistic Missile (ABM) Treaty (2002)
- U.S. Air Force (USAF) *Counterspace Operations* doctrine (2004)
- President George W. Bush's National Space Policy (2006)
- Operation Burnt Frost (2008)
- President Barack Obama's National Security Space Policy (2011)

---

[7] See, for example, Brad Roberts, *On Theories of Victory, Red and Blue*, Livermore, Calif.: Center for Global Security Research, Lawrence Livermore National Laboratory, No. 7, June 2020.

- Remarks of General William Shelton (Commander, USAF Space Command) regarding the Geosynchronous Space Situational Awareness Program (GSSAP) (2014).[8]

To complete a representative survey of Chinese-language national-level and military texts and defense periodicals, we leveraged several categories of Chinese primary sources. First, we identified eight key books published by various People's Liberation Army (PLA) institutions that reflect the most-authoritative available open-source discussions of Chinese military thinking on space, spanning 1987 to 2020.[9] We also included a 2021 book on U.S. space militarization, written by researchers at the PLA Strategic Support Force (PLASSF) Space Systems Department (SSD) Aerospace Engineering University (AEU), which is now the primary academic institution supporting China's operational space organization, PLASSF.[10] We also identified 18 journal articles on broad U.S. space activities and policies by PLA-affiliated authors to provide additional details and context for Chinese military thinking. Lastly, we supplemented these sources as necessary with articles that address specific U.S. space events that are not otherwise covered in the above sources. In total, these Chinese sources span decades of PRC space

---

[8] We acknowledge that the selection of these ten events introduces bias in the sense that this is *our* perception of events that may have influenced Russian or Chinese perceptions regarding U.S. activities in space. We have attempted to correct for this bias by identifying other events in the course of our native-language data collection efforts to allow for the possibility that *other* events have had a greater effect on Chinese or Russian perceptions.

[9] Academy of Military Science (AMS) [军事科学院], ed., *Science of Military Strategy* [战略学], 1st ed., Beijing: Academy of Military Science Press [军事科学出版社], 1987; AMS Military Strategy Department [军事科学院战略研究部], ed., *Science of Military Strategy* [战略学], 2nd ed., Beijing: Academy of Military Science Press [军事科学出版社], 2001; Chang Xianqi [常显奇], *Military Astronautics* [军事航天学], 2nd ed., Beijing: National Defense Industries Press [国防工业出版社], 2005; AMS Military Strategy Department [军事科学院战略研究部], ed., *Science of Military Strategy* [战略学], 3rd ed., Beijing: Academy of Military Science Press [军事科学出版社], 2013; Jiang Lianju [姜连举], ed., *Lectures on the Science of Space Operations* [空间作战学教程], Beijing: Military Science Press [军事科学出版社], 2013; Xiao Tianliang [肖天亮], ed., *Science of Military Strategy* [战略学], Beijing: National Defense University Press [国防大学出版社], 2015; Xiao Tianliang [肖天亮], ed., *Science of Military Strategy* [战略学], Beijing: National Defense University Press [国防大学出版社], 2017 revision; and Xiao Tianliang [肖天亮], ed., *Science of Military Strategy* [战略学], Beijing: National Defense University Press [国防大学出版社], 2020 revision.

For some recent Chinese views of space overall, see State Council Information Office of the People's Republic of China, *White Paper on China's Space Activities in 2016*, December 27, 2016; Ministry of Foreign Affairs (MFA) of the People's Republic of China, "Statements by Chinese Delegation at the 55th Session of the Scientific and Technical Subcommittee of the Committee on the Peaceful Uses of Outer Space," February 14, 2018; and Permanent Mission of the People's Republic of China to the United Nations, 2021.

[10] Feng Songjiang [丰松江] and Dong Zhenghong [董正宏], *Space, the Future Battlefield: New Situation and New Trends in the U.S. Militarization of Space* [太空未来战场: 美国太空军事化新态势新走向], Beijing: Current Affairs Press [时事出版社], 2021. For more on PLASSF's organizational structure, see Kenneth Allen and Mingzhi Chen, *The People's Liberation Army's 37 Academic Institutions*, Washington, D.C.: China Aerospace Studies Institute, 2020.

thinking, with the first published in 1984 and the most recent published in 2021, providing authoritative snapshots over time.[11]

For the analysis of Russian perceptions, we used the East View Information Services Online database to search Russian-language military and security periodicals to identify a publication that continually published research on space-related issues. We completed a broad search query using the terms "(*космо** OR *косми**) AND (*Amerika** OR *США*)."[12] The Russian newspaper *Krasnaia Zvezda* [*Red Star*]—the official newspaper of the Armed Forces of Russia, published by the Ministry of Defense—yielded the largest number of potentially relevant space-related articles (2,865).[13] With this baseline of articles, we reviewed and weeded out those articles that mentioned some U.S. space-related activity or policy only in passing to compile a final data set of relevant articles (199). Our analysis of these articles from *Red Star* was supplemented with high-level Russian strategic documents, including the Russian Federation's 2014 *Military Doctrine*, 2016 *Foreign Policy Concept*, and 2021 *National Security Strategy*. Finally, we included relevant articles from *Military Thought*—a monthly publication of military-theoretical articles by leading figures in Russia's Ministry of Defense, members of its General Staff, and other prominent uniformed and civilian thinkers.

It is important to understand what these Chinese and Russian sources can and cannot explain about perceptions regarding U.S. military space-related activities. These sources provide insights into internally focused Chinese and Russian analysis of U.S. space activities, largely separate from externally focused narratives, which are plausibly relatively more intended for perception-shaping. These sources are not necessarily self-reflective regarding their country's previous activities in space or how those actions might relate to U.S. space activities. Additionally, these sources do not necessarily specify Beijing's or Moscow's policy responses or planned future

[11] For other recent reviews of Chinese perceptions of U.S. space activities, see Hui Zhang, "The U.S. Weaponization of Space: Chinese Perspectives," presented at Nuclear Policy Research Institute Conference, *Full Spectrum Dominance: The Impending Weaponization of Space*, Warrenton, Va., May 16–17, 2005; Hui Zhang, "Chinese Perspectives on Space Weapons," in Pavel Podvig and Hui Zhang, eds., *Russian and Chinese Responses to U.S. Military Plans in Space*, Cambridge, Mass.: American Academy of Arts and Sciences, January 2008; Larry M. Wortzel, *The Chinese People's Liberation Army and Space Warfare*, Washington, D.C.: American Enterprise Institute, 2007; Erik Quam, *Examining China's Debate on Military Space Programs: Was the ASAT Test Really a Surprise?* Washington, D.C.: Nuclear Threat Initiative, January 31, 2007; Gregory Kulacki, *Anti-Satellite (ASAT) Technology in Chinese Open-Source Publications*, Cambridge, Mass.: Union of Concerned Scientists, July 1, 2009; Tong Zhao, *Trust-Building in the U.S.-Chinese Nuclear Relationship: Impact of Operational-Level Engagement*, dissertation, Atlanta, Ga.: Georgia Institute of Technology, 2014; and Kevin Pollpeter, Timothy Ditter, Anthony Miller, and Brian Waidelich, *China's Space Narrative: Examining the Portrayal of the US-China Space Relationship in Chinese Sources and Its Implications for the United States*, Montgomery, Ala.: China Aerospace Studies Institute, 2020.

[12] We continuously repeated this search throughout the project to add more-recent articles to the data set. The last update took place on September 15, 2021.

[13] For comparative reference, over this period, there were 188,174 articles published in the *Red Star* that were in the East View database. Currently, the newspaper is issued three times per week, but it has sporadically been issued more frequently, sometimes daily.

actions in response. Consequently, finding explicit evidence in these sources that would allow a causal argument to be made—for example, that U.S. activities drove specific Chinese and Russian behaviors in space—is difficult. Similarly, it is difficult to find explicit discussions in these sources of an intentional gap between internally focused perceptions and externally focused narratives about Chinese or Russian perceptions. We have attempted to address this lack of discussion by supplementing our analysis with other primary and secondary English-language sources to try to gain an initial sense of whether internally and externally focused messaging is inconsistent. But further studies to this end would be beneficial. Finally, the absence of explicit Chinese and Russian acknowledgment of a connection between their actions in space and those by the U.S. military does not mean that such a link does not exist. In the concluding chapter of the report, we have attempted to draw out some of these potential linkages using publicly available information.

As mentioned earlier, this report does not dwell on Chinese and Russian external narratives about U.S. space activities, since these have been explored elsewhere and are readily available in Chinese and Russian English-language publications. As one recent U.S. report on China's space narrative summarized, Beijing

> emphasizes China's role in promoting the peaceful uses of space, international space cooperation, and the advancement of humankind, while downplaying the national security aspects of its space program. In contrast, Chinese depictions of the U.S. space program portray the United States as emphasizing the military aspects of space, limiting international cooperation, restricting access to space, and working to destabilize humankind.[14]

This report is instead intended to explore how Chinese and Russian perceptions of U.S. military space-related activities, which we derive through our analysis of internally focused sources, have evolved over time.

Organization of This Report

In the next chapter, we briefly describe the ten events agreed upon with the sponsor that we leverage as analytic waypoints to understand how Chinese and Russian perceptions of U.S. activities in space may have evolved over time. For each event for which information is publicly available in one of our sources, we discuss China's and Russia's immediate reactions to the specific event. We also look for trends in perceptions regarding the "type" of event—i.e., reactions depending on whether the event represented a policy or doctrine change by the United States or reflected an action taken by the United States.

In the third and final chapter, we assess what these sources and trends suggest for current thinking regarding Chinese and Russian reactions to U.S. military activities related to space. We consider whether genuine Chinese and Russian concerns appear to exist regarding U.S. military

---

[14] Pollpeter et al., 2020, p. 20.

space-related activities or whether these concerns appear to be artificially exaggerated. We also analyze what the data suggest regarding the merits of the contention that China's and Russia's space activities are, at least in part, reactions to U.S. space activities. We conclude with a summary of the kinds of U.S. space-related activities that appear to have had the most significant effect on Chinese and Russian perceptions and actions within the space domain. Building on this analysis, we consider the potential implications of these findings, including what the next decade might hold in terms of Chinese and Russian activities in space. We conclude by assessing the U.S. public messaging, space developments, and in-space activities that might exacerbate or dampen these activities.

Chapter 2. Chinese and Russian Reactions to Pacing Events

International relations inherently represent a multiplayer, multi-move dynamic: Past interactions have the potential to affect present behavior, and current behavior can be affected by the potential for future interactions.[15] Chinese and Russian reactions to U.S. military activities in the space domain, therefore, can most usefully be examined in context—i.e., in relation to the broader history of actions and reactions. This insight led us to the methodology used to conduct this study.

Implementing this methodology, we proceed in this chapter as follows. For each of the ten U.S. military pacing events selected for inclusion in the study, we provide a brief overview and then present the data revealed in our examination of native-language Chinese and Russian sources, detailing immediate responses when such information exists. Responses consist of statements or specific actions for which the justification is specifically linked to a particular U.S. action. We conclude the chapter with a discussion of events that were not included in the initial list of ten. For context, Figures 2.1–2.4 at the end of the chapter provide an (admittedly imprecise) quantitative illustration of Chinese and Russian writing over time regarding these specific U.S. events and U.S. administrations.

Strategic Defense Initiative (1983) and U.S. Space Command Creation (1985)

Event Overview

In 1983, U.S. President Ronald Reagan announced his ambition to make nuclear weapons obsolete with a space-based missile defense program. The program envisioned multiple layers of futuristic technologies deployed in space that would identify and defeat an incoming large-scale barrage of nuclear weapons.[16] U.S. Space Command was established two years later, chiefly to manage the implementation of SDI, placing a single command in charge of the ballistic missile defense mission. U.S. Space Command responsibilities thus included tactical warning and space operations, control of space, directing space support activities, and planning for ballistic missile defense.[17]

---

[15] Duncan Snidal, "The Game Theory of International Politics," *World Politics*, Vol. 38, No. 1, October 1985, pp. 48–49.

[16] U.S. Department of State Archive, "Strategic Defense Initiative (SDI), 1983," webpage, undated.

[17] Edward J. Drea, Ronald H. Cole, Walter S. Poole, James F. Schnabel, Robert J. Watson, and Willard J. Webb, *History of the Unified Command Plan, 1946–2012*, Washington, D.C.: Joint History Office, Office of the Chairman of the Joint Chiefs of Staff, 2013, pp. 55–56.

China

SDI is a key part of the Chinese military's story of U.S. militarization of space and the space arms race, though such U.S. intent is often traced further back, to President John F. Kennedy.[18] SDI [战略防御倡议], also known as the Star Wars Program [星球大战计划], was directly discussed in all eight of the key PLA books and five of the 16 articles that we reviewed, and it was referenced as part of larger discussions of Reagan-era space policy in an additional two articles. This makes SDI the most referenced U.S. military activity in the space domain within the literature that we reviewed, both in terms of raw count and in span of relevancy to PRC perceptions.

Our literature sample indicates that PRC perceptions of SDI have evolved over time in response to the changing international political landscape. The earliest article, a seminal examination of SDI published in 1984 that was reportedly written in response to a request by the Chinese leadership and was thus highly influential in China's later response, interpreted SDI as the Reagan administration's abandonment of mutually assured destruction with the Soviet Union in favor of a sword-and-shield approach focused on ballistic missile defense and, notably, did not address any potential threat to China.[19] The author argued that the multilayered SDI ballistic missile defense shield freed the United States to wield its nuclear triad sword or, in other words, potentially enabled a first-strike U.S. nuclear posture. This, in turn, would counteract the Soviet Union's numerical superiority in intercontinental ballistic missiles (ICBMs) and enable the United States to negotiate from a position of strength.[20] Zhuang relates, however, that some U.S. scientists argued against SDI funding because the technical requirements were unachievably high.[21] Zhuang remains ambiguous as to whether he agrees with this argument, though later in the report he acknowledges that SDI has difficult technical hurdles to clear.[22] The continued focus on SDI, despite doubts over its technological feasibility, suggests greater focus on U.S.

---

[18] For earlier U.S. explorations of Chinese views of and responses to SDI, see Central Intelligence Agency (CIA), "Views of Chinese Military and Civilian Analysts on the Strategic Defense Initiative," January 1986, declassified May 27, 2011; Bonnie S. Glaser and Banning N. Garrett, "Chinese Perspectives on the Strategic Defense Initiative," *Problems with Communism*, Vol. 35, No. 2, March–April 1986, p. 28; John W. Garver, "China's Response to the Strategic Defense Initiative," *Asian Survey*, Vol. 26, No. 11, November 1986; John Wilson Lewis and Hua Di, "China's Ballistic Missile Programs: Technologies, Strategies, Goals," *International Security*, Vol. 17, No. 2, Fall 1992; and Brad Roberts, *China and Ballistic Missile Defense: 1955 to 2002 and Beyond*, Alexandria, Va., Institute for Defense Analyses, September 2003, pp. 11–18.

[19] Zhuang Qubing [庄去病], "Analysis of the U.S. 'Star Wars' Program" ["美国'星球大战'计划剖析"], *International Studies* [国际问题研究], April 1984, p. 28. For more on the article's importance, see Evan Medeiros, *Reluctant Restraint: The Evolution of China's Nonproliferation Policies and Practices, 1980–2004*, Stanford, Calif.: Stanford University Press, 2007, p. 259; and Alastair Iain Johnston, *China and Arms Control: Emerging Issues and Interests in the 1980s*, Ottawa, Canada: Canadian Centre for Arms Control and Disarmament, 1986.

[20] Zhuang, 1984, pp. 24–31, 36.

[21] Zhuang, 1984, p. 27.

[22] Zhuang, 1984, p. 27.

intent relative to U.S. capability. This focus on U.S. intent is also common in Russian analysis, in which the apparent assumption is that some technological progress will be made and, even if it is not, just the existence of such programs ultimately reveals U.S. intent that might be realized through other means down the road.

The 1987 edition of *Science of Military Strategy* (SMS), an authoritative source for PLA thinking during the period, similarly argued that the SDI program was intended to counter Soviet ICBM superiority. The authors concluded that the most important lesson of SDI was that strategic defense can be overcome through the development of new offensive capabilities. New offensive capabilities lead to the development of defensive capabilities, which results in a return to balance between offense and defense.[23] This aligns with Tong Zhao, a Chinese researcher at the Carnegie Endowment for International Peace in Beijing, who summarized Chinese views as follows: "The Reagan administration's SDI was not seen as mainly targeted against China, but the Chinese nuclear community's interpretation of the U.S. motivation behind SDI was an effort of the United States to regain strategic superiority over the Soviet Union."[24] More broadly, the authors of the 1987 SMS saw space warfare on the horizon, arguing that the United States was developing space warfare theories and space-based weapons in response to advancing technologies, and foresaw U.S. satellites being used in anti-satellite (ASAT) and anti-ICBM roles.[25]

Chinese authors writing in the early 2000s and 2010s, after the start of China's rise and the concomitant increase in political friction between the United States and the PRC, assume more-hostile intent on the part of Washington. For example, a 2004 article asserts that SDI was the initial U.S. step in seizing space control [夺制天权] as part of a broader national strategy to guarantee U.S. military and economic supremacy in outer space.[26] The authors of these later articles argue that SDI envisioned the placement of directed energy and kinetic weapons into orbit to both defend against Soviet ICBM strikes and coerce other nations into following the United States through the threat of blocking access to space.[27] The 2001 SMS shifted the Chinese

---

[23] AMS, 1987, p. 207.

[24] Tong, 2014, p. 214.

[25] AMS, 1987, p. 201.

[26] Zhu Tingchang [朱听昌] and Liu Jing [刘菁], "Fighting for Command of Space: The Development Process and Influence of the 'High Frontier' Strategy of the United States" ["争夺制天权:美国'高边疆'战略的发展历程及其影响"], *Military Historical Research* [军事历史研究], March 2004.

[27] One such capability was "Rods from God" [上帝之杖], the USAF idea to place bundles of telephone pole–sized tungsten rods in space that could be dropped on and destroy hardened targets without the fallout from nuclear weapons. See Zhu and Liu, 2004; Fan Gaoyue [樊高月] and Gong Xuping [宫旭平], "The Development and Evolution of U.S. Space Strategic Thought (Part 1)" ["美国太空战略思想的发展与演变(上)"], *National Defense* [国防], February 2016a, pp. 40–44; and Hu Guangzheng [胡光正] and Kan Nan [阚南], "To Seize the Power of Heaven, Who Is the Final Winner?" ["夺取制天权, 谁是最后的赢家?"], *Aerospace Knowledge* [航空知识], December 2006, pp. 9–11.

interpretation of SDI, arguing that it was an extension of the U.S. nuclear deterrence threat, and this view was echoed in the 2005 edition of *Military Astronautics*.[28] The authors readily cited such programs as SDI as evidence of U.S. ill intent. Yet they also ignored the implications of the programs' cancellation. This may imply a kind of confirmation bias on the part of PRC scholarship. If the assumption, especially after the dramatic change in U.S.-PRC military relations after the protests and government massacre in Tiananmen Square in 1989, is that the United States is a military threat, then it seems reasonable that PLA scholars would naturally gravitate toward data that support that assumption.

SDI remains a key historical waypoint for Beijing. Although SDI officially ended during the George H. W. Bush administration, the Chinese authors in our sample seemed to be in agreement that the U.S. Theater Missile Defense Program (TMDP) [战区导弹防御计划] and Ballistic Missile Defense Office (BMDO) [弹道导弹防御局] were continuations of SDI.[29] Beyond the military aspect, by the 2000s, as Chinese concerns over U.S. global efforts to undermine authoritarian regimes grew, Chinese strategists also viewed SDI as a key part of broader U.S. efforts to undermine the Soviet Union—and ultimately cause its collapse—through an expensive arms race across domains, from nuclear to space, along with other "covert" actions, such as ideological infiltration through "peaceful evolution."[30]

A related topic of interest to Chinese authors in our sample was the "High Frontier" [高边疆] series of private, nongovernmental reports advocating for SDI. In their view, as the first time that U.S. policy analysts openly argued that space should be considered a warfighting domain—the same as land, sea, and air—High Frontier marks the start of U.S. militarization of space.[31] Chinese analysts frequently argued that the High Frontier strategy recommended abandoning the previous U.S. strategies of mutually assured destruction and territorial balance theory, instead embracing the use of space technology to counteract perceived Soviet nuclear superiority and pursuing a new national strategy focused on U.S. space superiority.[32]

Authors in our sample also took interest in the establishment of U.S. Space Command [航天司令部航天司令部, 空间作战指挥部, or 美军航天]. The articles in our sample describe the development of U.S. Space Command as starting in the 1980s under Reagan-era space policies.

---

[28] Peng Guangqian [彭光谦] and Yao Youzhi [姚有志], *Science of Military Strategy*, Beijing: Military Science Publishing House [军事科学出版社], 2005 (translation of 2001 Chinese version), pp. 6–7; and Chang, 2005.

[29] Zhu and Liu, 2004.

[30] David Shambaugh, *China's Communist Party: Atrophy and Adaptation*, Berkeley, Calif.: University of California Press, 2008, p. 77. For PLA lessons learned on the U.S. role in the Soviet Union's ideological collapse, see the 2013 PLA National Defense University movie *Silent Contest* (较量无声). For a partial transcript, see "Silent Contest," transcript, Chinascope, March 5, 2014.

[31] AMS Military Strategy Department, 2001, p. 107; AMS Military Strategy Department, 2013, p. 183. For background on the reports, see Daniel O. Graham, *Confessions of a Cold Warrior*, Fairfax, Va.: Preview Press, 1995.

[32] Zhu and Liu, 2004, p. 118; and Fan and Gong, 2016, pp. 40–44.

The authors noted that while space was increasingly becoming important to the U.S. way of war, the space mission was diffused across the services.[33] They argued that U.S. Space Command became an integral part of the perceived militarization of space under the George W. Bush administration and that, over the 1990s and early 2000s, the space mission began to be concentrated within the USAF.[34] The Bush administration sought to counter a possible "Space Pearl Harbor" from a rising China by concentrating command, control, and planning in U.S. Space Command.[35] At the same time, the USAF was relying more on space power, which led to a shift away from pure aviation and a refocus on air and space.[36] One author argued that U.S. Space Command took responsibility for purported U.S. ASAT capabilities, including missiles, direct energy, and satellites.[37] All the authors agreed that the development of U.S. Space Command was yet another indicator of U.S. hegemonic designs toward space.

Writers in our sample frequently marked the beginning of the supposed U.S. quest for space hegemony from the point of a remark attributed to President Kennedy: "The fight for the hegemony of space is the core of the next ten years. The country that can control space controls the earth."[38] It is unclear from which of President Kennedy's public remarks this quote is taken, if any. No matter the provenance of the quote, authoritative Chinese texts argue it proves the United States has sought to control and militarize space from the beginning. The quote, whether directly cited or partially referenced, does not appear in the sources we reviewed that were published prior to 2001.[39] The quote may be an apocryphal rhetorical device that reinforces the idea that U.S. actions in space have always been militaristic.

Sources also cite the existence of military space programs, ASAT tests and other proposed offensive space weapons, and control of the moon for strategic purposes as elements of Kennedy-era space doctrine.[40] As the 2015 National Defense University (NDU) SMS argued,

> the U.S. Apollo program had, in addition to its stated purpose of landing on the moon, a secret mission to "capture" a secret Soviet test satellite . . . . [T]he United States took this opportunity [the space race] to usher in its era of space

---

[33] Hu and Kan, 2006, pp. 9–11; Chang, 2005.

[34] Hu and Kan, 2006, pp. 9–11.

[35] Fan and Gong, 2016a, pp. 40–44.

[36] Yuan Jun [袁俊], "U.S. Space Strategy and Space War Exercise" ["美军太空战略与太空战演习"], *China Aerospace* [中国航天], July 2001, pp. 26–29; and Fan and Gong, 2016a, pp. 40–44.

[37] Yuan, 2001, pp. 26–29.

[38] Yuan, 2001, pp. 26–29; Hu and Kan, 2006, pp. 9–11; AMS Military Strategy Department, 2001, p. 107; AMS Military Strategy Department, 2013, pp. 138–139; Jiang, 2013, pp. 1, 86; Xiao, 2015, p. 138.

[39] It is possible that the 2001 AMS SMS is the origin of this quote; at least, it is the first authoritative Chinese military source that we have found to contain the quote. See Peng and Yao, 2005, p. 100.

[40] Fan and Gong, 2016a, pp. 40–44; Yuan, 2001, pp. 26–29; Hu and Kan, 2006, pp. 9–11.

> hegemony and created an unprecedented "military hard power" and "strategic soft power" for the United States.[41]

President Kennedy's call for the peaceful development of space is completely absent from the Chinese sources we reviewed.[42]

Russia

Most space-related publications in Russian military journals refer to SDI as a starting point for the militarization of space and the space arms race, though the Soviet Union began work on its own orbital ASAT system (Istrebitel Sputnikov) in the early 1960s, itself a response to perceived U.S. militarization of space in the 1950s. For example, SDI is frequently cited as the foundation for supposed "space superiority" [*kosmicheskoye prevoskhodstvo*] efforts by future U.S. administrations.[43] SDI is identified as *the* essential program for the United States that would subsequently define the U.S. approach to space issues. Russian articles we reviewed emphasized that the United States lost the space race—i.e., the race for the first satellite and the first human in space—thus leading to the U.S. focus on gaining and maintaining space superiority in other facets.[44] A 1993 article, for example, argued that SDI was the first attempt to achieve superiority and undermine the strategic nuclear balance, asserting,

> it [SDI] was raised to the level of the priority task of state policy in Washington and initially provided for covering the entire American territory with an impenetrable anti-missile shield in order to be able to threaten the Soviet Union and the whole world with a strategic nuclear sword with impunity.[45]

Russian authors thus emphasized that SDI was an ill-conceived threat to global stability that would trigger an arms race.[46]

Red Star publications offer similar characterizations of SDI, regardless of the context. Although some Russian authors address the technical details and purpose of SDI and others focus on the implementation of SDI in the policy process,[47] the authors explicitly state that the

---

[41] Xiao, 2015, p. 139.

[42] John F. Kennedy, "Address Before the General Assembly of the United Nations," September 25, 1961.

[43] Andrey Shmygin, "Pul's Planety. Stavka na 'Blestyashchiye Kameshki i Glazki'" ["Pulse of the Planet. Betting on the 'Brilliant Pebbles and Eyes'"], *Krasnaia Zvezda* [*Red Star*], No. 108, 2001.

[44] Anatoliy Dokuchayev, "Rakety i Kosmos. Okhota Sredi Zvezd" ["Rockets and Space. Hunting Among the Stars"], *Krasnaia Zvezda* [*Red Star*], No. 264, 1999.

[45] Manki Ponomarev, "Voyennoye Obozreniye. SOI: Zhizn' Posle Smerti?'" ["Military Review. SDI: Life After Death?"], *Krasnaia Zvezda* [*Red Star*], No. 168, 1993b.

[46] Anna Polyakova, "Vyvod Oruzhiya v Kosmos Otkroyet Yashchik Pandory" ["Putting Weapons into Space Will Open Pandora's Box"], *Krasnaia Zvezda* [*Red Star*], No. 15, 2016; and Aleksey Lyashchenko, "Voyenno-Politicheskoye Obozreniye. Ambitsioznyye Plany Washingtona" ["Military-Political Review. Washington's Ambitious Plans"], *Krasnaia Zvezda* [*Red Star*], No. 105, 2002.

[47] "'Zvozdnyye Voyny': Kak SSSR Otvetil Reyganu" ["'Star Wars': How the USSR Responded to Reagan"], *Krasnaia Zvezda* [*Red Star*], No. 169, 2008.

initiative was technically unfeasible and unnecessary. In the articles we reviewed, the authors assert that President Reagan was naively persuaded by hawkish military advisers and an insatiable technical community eager for more resources. One source argued, "Reagan was a famous Hollywood actor, trade unionist, and effective governor of California. He had an unwavering faith in the omnipotence of the American technical genius, although he himself was a complete layman in the science and technology."[48] One article drew a parallel to the experience of the Union of Soviet Socialist Republics (USSR), noting that scientists would support politicians' technically infeasible ideas for a particular "wonder weapon" [*chudo-oruzhiye*] to gain funding for their research.[49]

Russian authors also used Western sources, and their criticisms of SDI, to justify their claims about its infeasibility. One 1993 article summarized a publication in *The Independent* that characterized SDI as a "deception" aimed at forcing the Soviet Union to enter an "arms race" and exhaust its financial resources.[50] Other authors suggested that the ideas were intentionally leaked to the Soviet Union as a form of psychological warfare to drive the USSR to spend itself into bankruptcy trying to recreate and counter what were transparently "bizarre" technologies.[51] This hypothesis was offered in several articles, which cited specific examples of technological failures (e.g., the Brilliant Pebbles program) to justify the point.[52] Moreover, Soviet scientists who conducted directed-energy experiments appear to have convinced the Kremlin that the threat from SDI was overstated.[53]

At the same time, while trying to downgrade the achievements of SDI, some articles, especially more-recent articles, acknowledge it was able to make some progress (although they emphasize that it was never fully implemented).[54] This recognition is then used to downplay subsequent U.S. space programs; the authors assert that a particular program offers nothing new

---

[48] "'Zvozdnyye Voyny': Kak SSSR Otvetil Reyganu" ["'Star Wars': How the USSR Responded to Reagan"], 2008.

[49] Manki Ponomarev, "Amerikanskikh Yadershchikov Po-Prezhnemu Zhdet Svetloye Budushcheye" ["American Nuclear Scientists Still Have a Bright Future"], *Krasnaia Zvezda* [*Red Star*], No. 174, August 4, 1992.

[50] "SOI: Bol'shaya Lozh' Reygana" ["SDI: Big Reagan's Deception"], *Krasnaia Zvezda* [*Red Star*], No. 238; 1993; and Dokuchayev, 1999.

[51] Mikhail Rebrov, "Sekrety Sekretnykh Sluzhb. DEZA, Ili Shirokiy Front 'Psikhologicheskoy Voyny'" ["Secrets of the Secret Services. DISINFO, or the Wide Frontier of 'Psychological Warfare'"], *Krasnaia Zvezda* [*Red Star*], 1995, p. 56; Manki Ponomarev, "SOI – Naslediye 'Kholodnoy Voyny'" ["SDI – A Legacy of the Cold War"], *Krasnaia Zvezda* [*Red Star*], No. 65, 1993a; and Mikhail Pogorelyy, "Ot Zvezd, Cherez Politicheskiye Ternii, - Na Zemlyu" ["From the Stars, Through Political Hardships, to the Earth"], *Krasnaia Zvezda* [*Red Star*], Nos. 108–109, 1993.

[52] Ponomarev, 1993a.

[53] Peter J. Westwick, "'Space-Strike Weapons' and the Soviet Response to SDI," *Diplomatic History*, Vol. 32, No. 5, November 2008.

[54] Vladimir Kuzar', "Voyenno-Politicheskoye Obozreniye. Cherez Preniya k Zvezdam" ["Military-Political Review. Through Debate to the Stars"], *Krasnaia Zvezda* [*Red Star*], No. 111, 2004.

compared with SDI.[55] The articles also emphasize that the Soviet Union successfully developed an "asymmetric response": That is, instead of engaging in a missile defense race, the Soviet Union deployed an ICBM (Topol-M) capable of defeating the U.S. missile defense system and thus ensuring nuclear retaliation.[56]

The authors in our sample also point to SDI's evolving objectives as proof that the concept was really seeking to gain advantage over Russia. The authors whose work we analyzed further note that while SDI was first characterized as a counter to Soviet missiles, it was subsequently justified as a collaborative effort to defend against "limited attacks" from "rogue" actors (i.e., terrorists) or accidental launches.[57] Generally, the tone of the articles talking about this shift of SDI objectives and openness to collaboration with Russia was one of skepticism. Most authors doubted there was even the slightest probability collaboration would come to fruition because it would require Russian access to SDI developments; defense, some authors argued, was a decoy, and Russia was still the main target.[58] This Russian focus on U.S. hostile intent toward Moscow appears to have shaped perceptions of SDI's technological feasibility, overlooking the otherwise obvious explanation that the change in the U.S. approach to SDI occurred because Washington recognized that the Reagan-era vision of SDI was technically infeasible.

President Clinton's National Space Policy (1996)

Event Overview

The Clinton administration's National Space Policy emphasized the peaceful exploration and use of space.[59] The five specific goals for the U.S. space program detailed in this strategy document were increasing knowledge through exploration, strengthening and maintaining U.S. national security, advancing technical competencies to improve economic competitiveness, encouraging investment in space technologies, and promoting international cooperation.[60]

---

[55] Aleksandr Khryapin, and Oleg Pyshnyy, "V 'Avangarde' Progressa" ["In the 'Vanguard' of Progress"], *Krasnaia Zvezda* [*Red Star*], No. 11, 2019.

[56] "'Zvezdnyye Voyny': Otvetit' Asimmetrichno" ["'Star Wars': Respond Asymmetrically"], *Krasnaia Zvezda* [*Red Star*], No. 164, 2007; and "'YEVRPRO': Asimmetrichnyy Otvet" ["European Missile Defense: Asymmetric Response"], *Krasnaia Zvezda* [*Red Star*], No. 222, November 30, 2011.

[57] "Programma SOI Stalkivayetsya s Ser'yeznymi Problemami" ["SDI Program Faces Major Challenges"], *Krasnaia Zvezda* [*Red Star*], No. 60, 1992.

[58] "Protivoraketnaya Oborona: Ukrepleniye Strategicheskoy Stabil'nosti Ili Novyy Vitok Gonki Vooruzheniy?" ["Missile Defense: Strengthening Strategic Stability or a New Round of the Arms Race?"], *Krasnaia Zvezda* [*Red Star*], No. 118, 2000.

[59] Note that the first U.S. National Space Policy was published by the Reagan administration in 1982, and subsequent administrations continued the practice.

[60] National Science and Technology Council, "Fact Sheet: National Space Policy," The White House, September 19, 1996.

China

The 1996 National Space Policy [国家航天政策 or 国家太空政策 or 国家空间战略] does not appear to be of great interest to Chinese analysts, though it is often associated with the U.S. strategy of space deterrence. The policy was not cited in the key publications we reviewed, and it is difficult to identify contemporaneous analysis. Outside this core sample, several more-recent articles, including some by PLA researchers, traced the U.S. national policy of space deterrence to the 1996 policy.[61] A 2017 article, by a nonmilitary Chinese academic whose views appear nonetheless to be representative of broader Chinese thinking, framed Clinton's policy as the beginning of continuity through the Bush and Obama administrations. All three "emphasized the importance and criticality of space for safeguarding national security, commercial and civil interests," in part through space deterrence, though, as will be discussed below, the Bush administration's preference for preemptive strikes was considered "very offensive," whereas the Obama administration was viewed as exercising "strategic restraint."[62]

However, broader U.S. military efforts during the Clinton administration were noted at the time. The 2001 AMS SMS contends,

> some countries are extending their "command of the air" theory to "command of the space" theory. . . . Emphasizing the increasing urgency and dependence of the U.S. on the outer space, the USAF requests the outer space be treated as a "strategic high ground for directing energy" [as in directed-energy weapons]. The U.S. military is also planning to develop space-based weapons and suggests the strategic concepts of "space control" and "space superiority."[63]

Similarly, the 2005 book *Military Astronautics* noted that the 1998 U.S. Space Command *Vision for 2020*[64] strategy identified the primary tasks of U.S. space power in the 21st century as "seizing control of space," "maintaining space dominance," and seeking to operationalize space

---

[61] For PLA-affiliated authors, see Zhou Lini [周黎妮], Fu Zhongli [傅中力], and Wang Shu [王姝], "Comparison Between Space Deterrence and Nuclear Deterrence" ["太空威慑与核威慑比较研究"], *National Defense Science and Technology* [国防科技], Vol. 36, No. 3, June 2015; Gaoyang Yuxi [高杨予兮], "The Historical Evolution of U.S. Space Deterrence Strategy" ["美国太空威慑战略的历史演进"], *International Study Reference* [国际研究参考], June 2017; Luo Xi [罗曦], "The Adjustments of U.S. Strategic Deterrence System and Their Implications to Sino-US Strategic Stability" ["美国战略威慑体系的调整与中美战略稳定性"], *Journal of International Relations* [国际关系研究], June 2017; and Gaoyang Yuxi [高杨予兮], "Adjustment of US Space Deterrence Strategy and Its Impact" ["美国太空威慑战略调整及其影响"], *Peace and Development* [和平与发展], March 2018, pp. 116–130, 135. For a non-PLA author, see He Qingsong [何奇松], "Fragile High Frontier: The Strategic Dilemma of U.S. Space Deterrence in the Post–Cold War Era" ["脆弱的高边疆:后冷战时代 美国太空威慑的战略困境"], *Social Sciences in China* [中国社会科学], April 2012.

[62] He Qingsong [何奇松], "Interactions Between China and the United States in the Area of Space Security" ["中美两国太空安全领域的互动"], *Journal of International Security Studies* [国际安全研究], September 2017.

[63] Peng and Yao, 2005, p. 124.

[64] U.S. Space Command, *Vision for 2020*, Colorado Springs, Colo.: Peterson Air Force Base, 1998.

warfare.[65] Other U.S. events noted in the book are the 1998 USAF Space Operations Doctrine, which provided guidance for space control, and comments by then–Secretary of Defense William Cohen about the importance of space deterrence.[66] Other analysts in our sample focused on the 1996 Global Positioning System (GPS) agreements signed during the Clinton administration, arguing that the Clinton administration had secured U.S. supremacy in the GPS market by allowing the U.S. military-industrial complex to sell GPS services to foreign entities, thus boxing out the nascent competition.[67] However, these Clinton administration–era U.S. space activities appear to have faded from prominence, as more-recent U.S. space activities were either more high-profile or more noteworthy to Chinese analysts.

Russia

In the collected articles, the 1996 U.S. National Space Policy is mentioned infrequently. The document is most often mentioned as a foil to demonstrate the shift in U.S. space policy from a peaceful space exploration orientation to a more aggressive posture emphasizing space superiority with the 2006 U.S. National Space Policy.[68] This suggests that Russian analysts were perhaps quietly acknowledging the more positive, more cooperative, and less confrontational aspects of the Clinton administration's policies but ultimately cast these aspects aside as an aberration within the overarching tendency of U.S. hostility in the space domain.

MIRACL Test (1997)

Event Overview

On October 17, 1997, the U.S. Army test-fired its ground-based MIRACL—a developmental program initiated under SDI—at a USAF satellite that the Pentagon stated had reached the end of its usable lifetime. U.S. officials said the purpose of the test was to investigate the effects of laser beams on satellite imaging sensors. The test marked the first time the United States had fired a laser at a satellite. While the test was described as an effort to understand the vulnerability of

---

[65] Chang, 2005, which states, in part:

> . . . proposed four major operational concepts for future space warfare: control of space, global engagement, full force integration, and global partnerships. This theory particularly emphasizes the necessity of making full use of integrated space force to carry out space-based ballistic missile defense and space-launched attack on all kinds of spacecraft, ballistic missiles, aircraft, warships, and high-value ground targets.

[66] Chang, 2005.

[67] Fan Gaoyue [樊高月] and Gong Xuping [宫旭平], "The Development and Evolution of U.S. Space Strategic Thought (Part 2)" ["美国太空战略思想的发展与演变(下)"], *National Defense* [国防], March 2016b, p. 53.

[68] Vladimir Kuzar', "Voyenno-Politicheskoye Obozreniye. Pole Boya – Vselennaya" ["Military-Political Review. The Universe Is the Battlefield"], *Krasnaia Zvezda* [*Red Star*], No. 60, 2006.

satellites to lasers, U.S. critics argued its real purpose was to demonstrate an offensive, ASAT capability.[69]

China

The MIRACL [中红外先进化学激光器] test is generally viewed in China as part of the U.S. development of ground-based laser counterspace capabilities for ASAT missions, though it was not specifically mentioned in any of the authoritative PLA sources we reviewed.[70] A 2009 article by a Chinese Academy of Science researcher described the MIRACL test as "showing that the U.S. military has the ability to use lasers to destroy enemy satellites" and claimed that, "since then, the laser has been secretly conducting satellite tracking and atmospheric correction tests at White Sands and Starfire testing areas" and that the United States has plans for deployment.[71] A 2013 AMS textbook on space operations, *Lectures on the Science of Space Operations*, does note that the MIRACL (not specifically the 1997 test) is one of several U.S. laser weapons that have "a certain operational capability."[72]

Russia

We did not find many mentions of this test in the data set. In the two articles that mention the test, it is characterized as being one of numerous examples that demonstrate the increasing desire of the United States to militarize space.[73] A few days after the test, however, the Russian foreign ministry issued a press release (in English) stating that laser programs "may become a step toward creating an anti-satellite potential."[74]

---

[69] "U.S. Test-Fires 'MIRACL' at Satellite Reigniting ASAT Weapons Debate," *Arms Control Today*, October 17, 1997; and "In Test, Military Hits Satellite Using a Laser," *New York Times*, October 21, 1997.

[70] The 2021 PLASSF SSD AEU book did mention the MIRACL test in passing. See Feng and Dong, 2021, p. 254. For two other Chinese defense industry articles that mention the general capability, see Li Yan [李焱], "The Latest Development Trend and Analysis of U.S. Space Weapons" ["美国太空武器最新发展动向及分析"], *Space International* [国际太空], 2008; and Wang Chaoqun [汪朝群], "Research on Space Defense" ["太空防御问题研究"], *Aerospace Electronic Warfare* [航天电子对抗], Vol. 25, No. 2, 2009.

[71] Zhang Jingxu [张景旭], "Progress in Foreign Ground-Based Optoelectronic Detecting System for Space Target Detection" ["国外地基光电系统空间目标探测的进展"], *Chinese Journal of Optics and Applied Optics* [中国光学与应用光学], Vol. 2, No. 1, February 2009.

[72] Jiang, 2013, p. 116.

[73] Kuzar', 2006; Vladimir Sidorov, "Boyevyye Lazery" ["Combat Lasers"], *Krasnaia Zvezda* [*Red Star*], No. 1, 2007.

[74] Press-Tsentr MID RF [Press Center of the Ministry of Foreign Affairs of the Russian Federation], "Briefing No 72," Integrum, October 21, 1997.

Commission to Assess United States National Security Space Management and Organization (2001)

Event Overview

In Section 1622 of the fiscal year 2000 National Defense Authorization Act,[75] the House Armed Services Committee directed the establishment of a committee to assess "the organization and management of space activities in support of U.S. national security."[76] The so-called Rumsfeld Commission found that the United States' relatively greater dependence on space made it asymmetrically vulnerable to attacks on systems located in space. Motivated by what the authors described as the potential for a "Space Pearl Harbor," the report recommended that the staffs of DoD and the CIA be reorganized to facilitate an increase in coordination and attention on the national security space program.[77] The commission's importance was further elevated by Donald Rumsfeld's role as Secretary of Defense for the incoming George W. Bush administration, since he was now in a position to turn these ideas into a reality.

China

The 2001 commission [美国国家安全空间管理和组织评估委员会] is seen as part of the Bush administration's realization that space is a military domain that is important for national security and must therefore be dominated, though it was not mentioned very often in the authoritative PLA books or broader PLA articles that we reviewed. *Military Astronautics* does mention the 2001 commission in passing in the context of discussing U.S. policy reports that reflect the belief in the United States that space has become a warfighting domain.[78] The book specifically cites the commission's report to say,

> Like land, sea and the atmosphere, space will become a battlefield. The United States must develop the capability to stop and defend itself against the enemy's military activities in and from space. . . . The national security of the United States is determined by its ability to operate successfully in space. If the United States wants to avoid a Pearl Harbor attack in space, it should seriously consider the possibility of launching effective attacks on space systems. Space power will become the main force on which the nation depends to safeguard national

---

[75] Public Law 106-65, National Defense Authorization Act for Fiscal Year 2000, October 5, 1999.

[76] Commission to Assess United States National Security Space Management and Organization, *Report of the Commission to Assess United States National Security Space Management and Organization*, Washington, D.C., January 11, 2001, p. vii.

[77] Commission to Assess United States National Security Space Management and Organization, 2001.

[78] Chang, 2005. For a similar reference, see Chen Jie [陈杰], Pan Feng [潘峰], and Su Tongling [苏同领], "EHF Satellite Communication System of the U.S. Army" ["美国天基太空监视系统"], *National Defense Technology* [国防科技], 2011.

security and implement military strategies. Therefore, the principal mission of the U.S. space power in the 21st century is to gain the upper hand in space.[79]

Russia

In the articles we sampled that mention the 2001 Rumsfeld Commission report, references to the report appear most frequently as a demonstration of the abrupt "shift" in U.S. space policy from the more peaceful, space exploration orientation of the Clinton administration to an era of "space superiority" pursued by the Bush administration. A 2006 article cites the report (along with the events on September 11, 2001) as triggering an inflection point in U.S. space policy.[80] Russian authors use quotes from the report and from Rumsfeld himself to illustrate the aggressive nature of the report's recommendations: for example, the language about seeking an "extraordinary military advantage" [*isklyuchitel'noye voyennoye preimushchestvo*] and "monopolistic" control of space.[81] Russian authors also keyed in on the report's warning of a "Space Pearl Harbor."[82] According to one author, these statements serve as proof that the United States is likely to start behaving more aggressively in space in the near future.[83] These articles also emphasize that the United States already actively utilizes existing capabilities to support conventional operations, heavily relying on space for navigation and reconnaissance.[84]

While the Rumsfeld Commission report featured prominently during our initial research stage, when a broad query was applied to the entire database of articles in East View, the number of articles appearing in *Red Star* that mentioned the report was relatively low. A similar trend held for *Military Thought*. Although several articles mention the report, it is only in passing, as a mechanism to introduce a discussion of new types of wars. These *Military Thought* articles have a similar sentiment to those in *Red Star*, using the same quotes regarding a "Space Pearl Harbor" and stressing that increased U.S. dependence on space creates an exploitable vulnerability.[85]

---

[79] Chang, 2005.

[80] Kuzar', 2006.

[81] Seumas Mil, "Bez Kommentariyev. Mozhno Li Pobedit' Soyedinennyye Shtaty?" ["No Comments. Can the United States Be Defeated?"], *Krasnaia Zvezda* [*Red Star*], No. 35, February 22, 2002; and Aleksey Ventslovskiy, "V Pogone za Zvezdami" ["In Pursuit of the Stars"], Krasnaia Zvezda [*Red Star*], No. 211, 2004.

[82] Kuzar', 2006.

[83] Ventslovskiy, 2004.

[84] Ventslovskiy, 2004.

[85] S. N. Konopatov and Ye. A. Starozhuk, "Kosmicheskiye Sistemy v Novoy Srede Bezopasnosti" ["Space Systems in a New Security Environment"], *Voennaia Mysl'* [*Military Thought*], No. 1, January 2019.

Withdrawal from the Anti-Ballistic Missile Treaty (2002)

Event Overview

The United States formally withdrew from the 1972 ABM Treaty on June 13, 2002, having proclaimed its plan to withdraw from the treaty six months prior. The ABM Treaty had prohibited the deployment of nationwide missile defense systems by the Soviet Union and United States.[86]

China

China has always been concerned by U.S. ballistic missile defenses, and the withdrawal from the ABM Treaty [反弹道导弹条约 or 反导条约] is generally viewed as an extension of SDI's intent to implement missile defense to enable U.S. offensive operations.[87] *Military Astronautics* argued it "[paved] the way for building and deploying the missile defense system. This further indicated the United States' determination to get prepared for full space warfare."[88] The 2013 AMS book *Lectures on the Science of Space Operations* similarly states,

> the pace at which [the United States] has enhanced the ability of its space operations for actual warfare has continually accelerated, continually widening the gap of its advantages over other nations; the situation is quite compelling. First, the United States withdrew from the Anti-Ballistic Missile Treaty in 2001, clearing away an obstacle to the National Missile Defense system and to space operations, and it formally deployed a missile defense system, thus comprehensively launching its space operations plans.[89]

While the treaty withdrawal was not addressed in other key PLA books reviewed, such as the 2013 AMS SMS or the 2020 NDU SMS, it was discussed in some of the journal articles that we reviewed. These articles argue the George W. Bush administration withdrew because the treaty limited the deployment of high-altitude interceptors being developed by the SDI-legacy TMDP and the BMDO.[90]

Beijing's response to Washington's withdrawal from the ABM Treaty is one example of an evolving Chinese public discourse on U.S. space activities. At first, when the Bush

---

[86] Wade Boese, "U.S. Withdraws from ABM Treaty; Global Response Muted," *Arms Control Today*, July/August 2002.

[87] For a contemporary review of Chinese reactions to the ABM Treaty withdrawal, see Roberts, 2003. For a key Chinese review of U.S. thinking on the withdrawal, see Wang Aijun [王爱娟] and Shi Bin [石斌], "Strategic Debate on the Issues Concerning the Treaty on the Limitation of Anti-Ballistic Missile Systems (ABM) and Missile Defense System Within the U.S. Government (1983–2001)" ["美国政府内部围绕反导条约与导弹防御体系问题的战略论争 (1983–2001)"], *China Military Science* [中国军事科学], January 2015.

[88] Chang, 2005.

[89] Jiang, 2013, p. 23.

[90] Fan and Gong, 2016a, pp. 40–44; Zhu and Liu, 2004.

administration withdrew, China was relatively quiet, apparently because it saw the move as an expected fulfillment of U.S. strategy.[91] In its 2002 defense white paper, China struck a relatively conciliatory tone of understanding Washington's motivations but calling for restraint on missile defense:

> China's stand on the issue of missile defense is consistent and clear-cut. China understands the relevant countries' concern over the proliferation of weapons of mass destruction (WMD) and their means of delivery. But, like many other countries, China holds that this issue should be resolved through political and diplomatic means, with the common efforts of the international community. China regrets the abrogation of the Anti-Ballistic Missile Treaty. . . . China hopes that the relevant countries will heed the opinions of the international community, and act prudently on the issue of missile defense.[92]

However, Beijing now much more openly criticizes the withdrawal as part of relentless U.S. efforts to develop offensive missile operations. As an MFA spokesperson said in October 2021, "As is known to all, the U.S. withdrew from the Treaty on the Limitation of Anti-Ballistic Missile Systems and the Intermediate-Range Nuclear Forces Treaty, and constantly advances the deployment of a global anti-ballistic missile system."[93] Chinese officials have also contended that the X-Band radar of the Terminal High Altitude Area Defense (THAAD) anti-ballistic missile system deployed in the Republic of Korea supports the monitoring of Chinese missile tests that can enhance U.S. strategic ballistic missile defense capabilities, discussed more in Chapter 3.

Russia

Mentions of the U.S. withdrawal from the ABM Treaty are common in *Red Star*. Authors often use the event to list U.S. transgressions in space or within a broader discussion of Russia's special bilateral great-power relationship with the United States.[94] Well before President Bush announced the U.S. decision to withdraw from the treaty, articles in *Red Star* started sounding the alarm during President Clinton's first term about what they viewed as an imminent U.S. decision to refuse to comply with the treaty. The articles repeatedly criticized the prospective move and put forward suggestions for how to "save" the treaty.[95]

---

[91] Roberts, 2003, pp. 32–34.

[92] PRC Information Office of the State Council, "China's National Defense in 2002," Xinhua, December 9, 2002.

[93] MFA of the People's Republic of China, "Foreign Ministry Spokesperson Wang Wenbin's Regular Press Conference on October 19, 2021," October 19, 2021a.

[94] Our data set of 199 articles excluded those articles that simply use withdrawal from the treaty as a reference point and focused solely on those articles that provide more context and relevant discussion.

[95] Viktor Dontsov, "'Tri Plyus Tri.' Pod Takim Nezateylivym Nazvaniyem Skryvayetsya Amerikanskaya Programma Sozdaniya Natsional'noy PRO" ["'Three plus Three.' The American Program for the Creation of a National Missile Defense System Is Hidden Under This Unpretentious Name"], *Krasnaia Zvezda* [*Red Star*], No. 260, 1998.

The overall tone of the publications in our sample that discussed the U.S. presumptive decision to withdraw from the treaty was very negative, with authors emphasizing that the move was a threat to strategic stability that did not serve the U.S. stated purpose for abrogating the treaty.[96] For example, the tone and substance reflected skepticism regarding the claim that the missile defense system was not directed at Russia. Some authors tried to find technical evidence to demonstrate that alternative solutions would provide better, more affordable protection from the "intractable" threat posed by rogue countries.[97] Other articles argued rogue states lacked the technical competence to produce missiles capable of reaching the United States.[98] And other articles suggested that the threat to Russia, if there was a missile threat, was greater than the threat to the United States. Thus, the authors suggested that, if a global or European missile defense system were to be built, it should be done jointly with Russia.[99]

According to Russian authors, the United States correctly pivoted to the main threat pervading the international landscape: "third world" or "rogue" actors.[100] Yet, the authors argued, the withdrawal from the treaty would only exacerbate the situation, expanding the threat surface for the United States by creating a need for the United States to defend itself from missiles that otherwise would have been limited by the treaty—i.e., Russian missiles.[101] The logic was that withdrawal from the treaty would start an arms race, with Russia deploying more offensive weapons to offset U.S. defenses. This stands in contrast to the general Chinese perspective that downplays the third-party threat (Russia, North Korea, etc.) to the United States. Moreover, the authors argue such an action would initiate the next spiral of the arms race, forcing other countries to develop weapons capable of penetrating such defense and expanding the arms race into the space domain.[102] Consistent with the experience of SDI, we also found

---

[96] Nikolay Mikhaylov, "K Voprosu o Dogovore po PRO 1972 g" ["On the 1972 ABM Treaty"], *Krasnaia Zvezda* [*Red Star*], No 273, December 30, 1999. Note that the same narrative dominates discussions of this event in articles in *Military Thought*. See, for example, O. I. Antsupov and A. S. Zhikharev, "Analiz Osnovnykh Kontseptual'nykh Podkhodov k Sozdaniyu Sistem Strategicheskoy PRO SSHA i Rossiyskoy Federatsii" ["Analysis of the Main Conceptual Approaches to the Creation of Strategic Missile Defense Systems of the USA and the Russian Federation"], *Voennaia Mysl'* [*Military Thought*], No. 6, 2015.

[97] Antsupov and Zhikharev, 2015; and Dmitriy Andreev, "Gospodstvuyushchiye Vysoty Kosmicheskikh Voysk" ["Superior Heights of the Space Forces"], *Krasnaia Zvezda* [*Red Star*], No. 14, 2008.

[98] Mikhail Falaleyev, "Ministr Oborony Rossiyskoy Federatsii Sergey Ivanov: Glavnyy Kriteriy - Bezopasnost' Rossii" ["Defense Minister of the Russian Federation Sergei Ivamov: The Main Criterion Is the Security of Russia"], *Krasnaia Zvezda* [*Red Star*], No. 136, 2001.

[99] "Protivoraketnyy Shchit Dlya Yevropy" ["Missile Defense Shield for Europe"], *Krasnaia Zvezda* [*Red Star*], No. 222, 2006.

[100] "Protivoraketnaya Oborona: Ukrepleniye Strategicheskoy Stabil'nosti Ili Novyy Vitok Gonki Vooruzheniy?" ["Missile Defense: Strengthening Strategic Stability or a New Round of the Arms Race?"], 2000.

[101] "Protivoraketnaya Oborona: Ukrepleniye Strategicheskoy Stabil'nosti Ili Novyy Vitok Gonki Vooruzheniy?" ["Missile Defense: Strengthening Strategic Stability or a New Round of the Arms Race?"], 2000.

[102] Oleg Falichev, "Vizit. Ministr Oborony Rossii v Belgrade" ["Visit. Russian Defense Minister in Belgrade"], *Krasnaia Zvezda* [*Red Star*], No. 270, 1999.

that the authors used domestic U.S. sources to buttress their critiques of the U.S. decision to withdraw from the treaty. The authors point to a U.S. government captive to the military-industrial complex, as well as partisan debates about the decision for reelection purposes.[103]

On the day of the withdrawal, the response from the Kremlin was limited to an announcement that Russia would no longer be bound by the Strategic Arms Reduction Treaty II, a treaty that had never entered into force.[104] Senior Russian officials publicly emphasized the embryonic status of U.S. missile defense efforts as justification for Russia's muted response. For example, Russian Defense Minister Sergei Ivanov described U.S. missile defense capabilities as being "virtual" and therefore requiring no immediate response.[105] Prior to the United States formally withdrawing from the treaty, articles in the data set discussed possible Russian responses. The authors cited Russia's technical prowess and status as the first country to successfully test a missile defense system in 1961, more than two decades before the United States was able to field such a system. They thus judged that Russia possessed the necessary intellectual, military, and technical capacity to retaliate asymmetrically, using inexpensive and effective means.[106]

Russian authors argued the Soviet Union had already demonstrated it would prioritize strategic stability over attempting to limit vulnerability to missile strikes.[107] The authors emphasized that the USSR had been in the lead in building missile defenses, but, to avoid starting an arms race, it had made the controversial decision to give up its lead and sign the ABM Treaty in 1972. One article notes,

> the 1972 ABM Treaty was a major departure from the traditional Soviet position. For many years, almost until 1972, both Soviet external messaging and official statements treated the idea of refusing to deploy missile defense only as an insidious plan of American imperialism, aimed at depriving the USSR of the opportunity to defend itself from a nuclear missile attack.[108]

---

[103] Sergey Ryzhkov, "Chto Budet s Dogovorom Po Otkrytomu Nebu?" ["What Will Happen to the Treaty on Open Skies?"], *Krasnaia Zvezda* [*Red Star*], No. 77, 2020; and Aleksandr Gol'ts, "PRO Natselena... v Klintona" ["Missile Defense Targets... Clinton"], *Krasnaia Zvezda* [*Red Star*], No. 146, 1996. The authors argue that both the Democratic and Republican parties desire a defensive system, the only difference being timing and scope.

[104] Press-Tsentr MID RF [Press Center of the Ministry of Foreign Affairs of the Russian Federation], "O Pravovom Statuse Dogovora Mezhdu Rossiyey i SSHA o Dal'neyshemsokrashchenii i Ogranichenii Strategicheskikh Nastupatel'nykh Vooruzheniy" ["On the Legal Status of the Treaty Between Russia and the United States on the Further Reduction and Limitation of Strategic Offensive Arms"], June 14, 2002.

[105] Sergey Sokut, "Uroki. Usloviya Prodiktoval Vashington" ["Lessons Learned. The Terms Were Dictated by Washington"], *Nezavisimoe Voennoe Obozrenie* [*Independent Military Review*], June 21, 2002, p. 20.

[106] Falaleyev, 2001.

[107] Anatoliy Dokuchayev, "Tochku Vstrechi Izmenit' Nel'zya. Zavtra Protivoraketchiki Rossii Otmetyat Yubiley" ["The Meeting Point Cannot Be Changed. Tomorrow, Russia's Missile Defense Forces Will Celebrate Their Anniversary"], *Krasnaia Zvezda* [*Red Star*], No. 42, 2001.

[108] V. N. Tsygichko and A. A. Piontkovskiy, "Dogovor Po PRO: Nastoyashcheye i Budushcheye" ["ABM Treaty: Present and Future"], *Voennaia Mysl'* [*Military Thought*], No. 1, 2000.

This event also represents one of the few instances in which a causal relationship can be drawn between a U.S. action and a Russian counteraction. Although financial hardships halted Russia's ability to develop missile defense in the 1990s, with the U.S. withdrawal from the ABM Treaty, scarce resources were pushed back into these programs.[109] Withdrawal from the treaty, according to Russian authors, also created the requirement for many new types of weapons, including ASAT weapons and munitions that could be placed in space.[110] Moreover, given the pending militarization of space, Russian authors argued Russia now had to "create a unified aerospace defense system using weapons based on new physical principles."[111]

USAF *Counterspace Operations* Doctrine (2004)

Event Overview

Following on the heels of the 2001 *Quadrennial Defense Review Report*, which called for the United States to exploit the advantages of the ultimate high ground of space,[112] the USAF published its *Counterspace Operations* doctrine.[113] The document stated that counterspace operations were critical to prevailing in a conflict and that the United States required "space superiority" to ensure freedom to operate in the domain while denying an adversary's ability to utilize space. The new doctrine thus addressed planning for space situational awareness; defensive operations to protect U.S. satellites and spacecraft; and offensive operations to deceive, disrupt, deny, degrade, and destroy adversary space capabilities.[114]

---

[109] Aleksey Martymyanov and Aleksandr Dolinin, "Nevidannyy Proryv" ["The Unseen Breakthrough"], *Krasnaia Zvezda* [*Red Star*], No. 239, 2006.

[110] Sergey Pechurov, "Formuly Akademika Kokoshina, Ili Kak Obespechit' Intellektual'noye Prevoskhodstvo Rossii" ["Formulas by Academician Kokoshin, or How to Ensure Russia's Intellectual Superiority"], *Krasnaia Zvezda* [*Red Star*], No. 69, 2014; Vladislav Runov, "Ne Narushat' Strategicheskoy Stabil'nosti" ["Do Not Disrupt Strategic Stability"], *Krasnaia Zvezda* [*Red Star*], No. 86, 2001; Yaroslav Yastrebov, "Zametki Obozrevatelya. Vzaimodeystviye Radi Strategicheskoy Stabil'nosti" ["Notes of an Observer. Collaboration for the Strategic Stability"], *Krasnaia Zvezda* [*Red Star*], No. 239, 2001; and M. G. Valeev, A. V. Platonov, and V. I. Yaroshevsky, "O Krizisakh vo Vzaimodeystvii Rossii i SSHA v Oblasti Protivoraketnoy Oborony" ["On Crises in the Interaction of Russia and the United States in the Field of Missile Defense"], *Voennaia Mysl'* [*Military Thought*], No. 7, July 2021.

[111] A. V. Supryaga, "O Voynakh XXI Veka" ["On Wars of the 21st Century"], *Voennaia Mysl'* [*Military Thought*], No. 6, 2002.

[112] DoD, *Quadrennial Defense Review Report*, Washington, D.C., September 30, 2001.

[113] Air Force Doctrine Document 2-2.1, *Counterspace Operations*, Washington, D.C.: Headquarters Air Force Doctrine Center, August 2, 2004.

[114] Air Force Doctrine Document 2-2.1, 2004.

China

The 2004 USAF *Counterspace Operations* doctrine [反太空行动方针] is generally viewed as another step toward enabling U.S. military operations in space, though it is not commonly cited in authoritative PLA writings. The 2013 AMS book *Lectures on the Science of Space Operations* frames the document as only part of a series of USAF doctrines that detail how the United States plans to use space for military operations, noting that the 2004 doctrine "proposed the concept of 'space campaigns,' clarifying the command and control and the guiding principles of defensive space confrontational actions and offensive space confrontational actions, as well as the requirements and procedures for carrying out joint space campaigns."[115] In addition to key authoritative texts, one 2010 PLA article, by authors at what is now PLASSF SSD AEU, framed the 2004 doctrine as an abrupt—but, in retrospect, unsurprising—shift in U.S. space strategy to focus more on space situational awareness.[116] Separately, a 2006 *People's Daily* article by Teng Jianqun, a former PLA officer, framed the 2004 document as evidence that the United States is militarizing space, to such an extent that each military service has its own space doctrine.[117]

Russia

The USAF *Counterspace Operations* doctrine is frequently mentioned in the data set, yet it is rarely discussed by itself. Rather, the doctrine is mentioned in connection with SDI and the Rumsfeld Commission report. The compilation of these events reveals what Russian writers view as the U.S. ambition to add "space superiority" or the "space echelon" [*kosmicheskiy eshelon*] to the existing U.S. superiority in land- and sea-based echelons.[118] The articles reviewed thus portray the doctrine as an illustration of the U.S. shift to a more hawkish approach to space exploration, which culminated in an active attempt to militarize space. Emphasizing the doctrine's central concept of realizing space superiority, Russian authors argue this concept demonstrates that the United States seeks space primacy—freedom from attack and freedom to attack.[119]

---

[115] Jiang, 2013, p. 13.

[116] Su Xiancheng [于小红], Yu Xiaohong [刘震鑫], and Liu Zhenxin [苏宪程], "Analysis of Development of U.S. Space Situation Awareness" ["美国空间态势感知发展分析"] *Journal of the Academy of Equipment Command & Technology* [装备指挥技术学院学报], April 2010.

[117] Teng Jianqun [滕建群], "Don't Drive War into Space" ["莫将战车开进太空"], *People's Daily*, November 4, 2006. In 2006, Teng was Deputy Secretary General of China Arms Control and Disarmament Association, a government think tank, and he then moved to the MFA think tank, China Institute of International Studies.

[118] "'Zvezdnyye Voyny': Otvetit' Asimmetrichno" ["'Star Wars': Respond Asymmetrically"], 2007.

[119] Vladimir Sidorov, "Pentagon Rvetsya v Kosmos" ["Pentagon Rushes into Space]", *Krasnaia Zvezda* [*Red Star*], No. 46, 2009.

President Bush's National Space Policy (2006)

Event Overview

The George W. Bush administration's National Space Policy emphasized strengthening U.S. leadership in space to ensure the availability of capabilities to support U.S. national security and foreign policy objectives and to enable "unhindered U.S. operations in and through space to defend our interests there."[120] The policy rejected future arms-control agreements, to the degree that such agreements might impinge on the ability of the United States to leverage space to meet its national security needs. It also emphasized the importance of freedom of action in space and asserted the U.S. right to deny access to space to any actor "hostile to U.S. national interests."[121]

China

The Chinese authors in our sample universally interpreted the 2006 U.S. National Space Policy as further proof of the U.S. desire to control and militarize space and, more broadly, of the Bush administration's ambitions for space dominance. For example, PLA authors wrote in 2014 that the policy

> expressed the Bush administration's hardline stance of vainly seeking to dominate space, treating space as its own unique territory, and elevating purposeful interference in its space system [by a foreign actor] to be equal with infringement of U.S. sovereignty. In addition, the Bush administration's national space policy also proposes to openly discuss the possibility of destroying other countries' satellites. The Bush's space strategy has brought the unilateralism and hegemonic thinking to the fullest.[122]

Chinese authors argued the 2006 policy's inclusion of space in U.S. sovereignty effectively folded the heavens into U.S. national defense interests. The Chinese authors noted that the United States had effectively weaponized space by including space in the U.S. strategic defense plan.[123] *Strategic Air Force*, a quasi-official 2009 book by the PLA Air Force (PLAAF) that, in part, advocated for a PLAAF role in Chinese space operations, said,

> The new policy clearly focuses on space security and space weapon development, and requires the United States to "develop and maintain U.S. superiority" . . . . The new [ballistic missile defense] policy emphasizes the need to ensure a strong space force to effectively protect America's space interests:

---

[120] The White House, "U.S. National Space Policy," August 31, 2006, p. 2.

[121] The White House, 2006, p. 2.

[122] Gaoyang Yuxi [高杨予兮] and Ke Long [柯隆], "New Trends of U.S. Space Cooperation Policy" ["美国太空合作政策新动向"], *International Study Reference* [国际研究参考], June 2014, pp. 1–5, 45.

[123] Gaoyang and Ke, 2014, pp. 1–5, 45; Fan and Gong, 2016b, p. 54; and Peng Huiqiong [彭辉琼], Lv Jiuming [吕久明], and Lu Jiangong [路建功], "Analysis of the Main Results of American Space Combat Exercises" ["美国太空作战演习主要成果探析"], *Aerospace Electronic Countermeasures* [航天电子对抗], Vol. 35, No. 2, 2019.

> The United States must not be obstructed when developing space weapons and performing space missions.[124]

One 2007 article even argued the 2006 policy proposed developing space weapons as one way to support counterterrorism in the post-9/11 era.[125] The 2021 PLASSF SSD AEU book similarly remarks that the 2006 policy, for the first time at the national level, put "space dominance" at the same level of importance as "air dominance" and "sea dominance."[126]

Beyond the 2006 policy, other remarks by Bush administration officials reinforced this view. In addition to citing the 2006 U.S. National Space Policy, one set of authors quoted a U.S. Secretary of the Air Force as saying, "Weapons will appear in space, and we should be in a leading position in this process."[127] The authors argued that, when combined with other U.S. actions taken during the Bush era, such as the withdrawal from the ABM Treaty, the 2006 policy made clear that the United States had begun to militarize space. Chinese analysts expressed the view that the Bush administration's space policy marked Washington's transparent desire to extend its purported space hegemony [太空霸权]. Chinese analysts also cited a comment reportedly made by National Reconnaissance Office (NRO) Director Peter Teets in 2004 that the United States "lacks the capability to dominate space. We are not a space hegemon . . . in fact, we need to achieve this goal."[128]

Russia

The articles in the data set that mention this event focus on the prominent role of the space superiority concept in the 2006 U.S. National Space Policy. The articles emphasize that the Pentagon's rhetoric made U.S. plans and intentions with respect to the space domain very explicit, citing such U.S. statements as "who owns the space, owns the world."[129] Moreover, the articles in *Red Star* stress that the space superiority concept was not limited to declaratory policy, observing that operationalizing the concept received substantial funding and arguing the United States was prepared to spend as much as necessary to militarize space.[130]

---

[124] Zhu Hui [朱晖], ed., *Strategic Air Force* [战略空军论], Beijing: Blue Sky Press [蓝天出版社], 2009, p. 325.

[125] Hu Xujie [胡绪杰] and Zhang Zhifeng [张志峰], "Research on the Development of U.S. Aerospace Forces" ["美国航天力量发展研究"], *O. I. Automation* [兵工自动化], Vol. 26, No. 9, 2007.

[126] Feng and Dong, 2021, p. 122.

[127] It is unclear which U.S. Secretary of the Air Force or speech the authors are referring to, as no citation is given. See Hu and Kan, 2006, pp. 9–11; and Zhu and Liu, 2004.

[128] Xiao, 2015, pp. 140–141. The book misidentified Teets as the director of the U.S. National Security Agency (NSA, 国家安全局局长).

[129] Aleksey Lyashchenko, "Voyenno-Politicheskoye Obozreniye. Reanimatsiya Gonki Yadernykh Vooruzheniy" ["Military-Political Review. Reanimation of the Nuclear Arms Race"], *Krasnaia Zvezda* [*Red Star*], No. 147, 2005.

[130] "Kosmicheskiy Machizm Opasen Dlya Chelovechestva" ["Space Machismo Is Dangerous for the Humanity"], *Krasnaia Zvezda* [*Red Star*], No. 23, 2007; and Aleksey Ventslovskiy, "Amerike Nuzhny Den'gi" ["America Needs

Articles from *Military Thought* describe the 2006 U.S. National Space Policy along similar lines, emphasizing U.S. statements about the monopolistic use of space. These articles heavily criticize the United States, detailing the perception that the United States is trying to assume the roles of both judge and executioner should it assess that another country's activities could potentially harm U.S. interests. The authors further criticize the United States for asserting that it should be entirely free to act as it desires in space while declaring that the actions of other countries to protect their own interests in space constitute an attempt to hinder U.S. space interests, giving the United States the right to act.[131]

Consistent with Russian perceptions of SDI, *Red Star* authors again point to U.S. critiques of the Bush space program. For example, some authors cite the domestic critique that the United States would be better served spending this money on education and health care.[132]

In general, the tone of the articles in our sample that describe U.S. space policy under the Bush administration reads as ambivalent. Yet a rather delicate rhetorical balance is struck. On the one hand, the authors stress the threat from the developments described above, especially for the countries without advanced space capabilities. They emphasize that Russia would prefer not to have a space race at all and accuse the United States of being uncooperative and provocative in its space activities.[133] The prevailing narrative in these articles is that the Bush administration's pursuit of a monopoly in space and space superiority undermines global stability. U.S. efforts have forced Russia's hand: It must step in, serving as the peaceful guardian of space. The authors emphasize Russian efforts to prevent the militarization of space through strengthened international regulations, arguing that while Russia has worked jointly toward this end with other partners, the United States and some of its allies regularly impede such efforts.[134]

On the other hand, even while characterizing U.S. actions as a threat that Russia is obligated to counter, the authors are careful not to downgrade Russia's capabilities. This balancing act usually takes the form of stating that while the United States was leading the space race (and

---

Money"], *Krasnaia Zvezda* [*Red Star*], No. 21, 2005. The articles in the *Red Star* also emphasize the specific weapons that were perceived as being researched and developed at the time, including Rods from God, "kamikaze" microsatellites, and Prompt Global Strike. See, for example, "Kosmicheskiy Machizm Opasen Dlya Chelovechestva" ["Space Machismo Is Dangerous for the Humanity"], 2007; Vladimir Kuzar', "Kosmos Nadevayet Kamuflyazh" ["The Space Puts on the Camouflage"], *Krasnaia Zvezda* [*Red Star*], No. 37, 2008a; and V. Satarov, "Stremitel'nyy Global'nyy Udar" ["Prompt Global Strike"], *Krasnaia Zvezda* [*Red Star*], No. 168, 2007.

[131] B. F. Chel'tsov, "Voprosy Vozdushno-Kosmicheskoy Oborony v Voyennoy Doktrine" ["Aerospace Defense Issues in the Russian Military Doctrine"], *Voennaia Mysl'* [*Military Thought*], No. 4, April 2007.

[132] Vadim Markushin, "Marsiada Busha" ["Bush's Martian Project"], *Krasnaia Zvezda* [*Red Star*], No. 4, 2004.

[133] Dmitriy Kirsanov, "Kosmicheskiye Orbity Pentagona" ["Pentagon Space Orbits"], *Krasnaia Zvezda* [*Red Star*], No. 66, 2012.

[134] Vladimir Kuzar', "SSHA: Problemy Rastut, Ambitsii Te Zhe" ["The USA: Growing Problems, Same Ambitions"], *Krasnaia Zvezda* [*Red Star*], No. 15, 2008b.

creating a threat), other countries were rapidly catching up and closing the capability gap.[135] Russian authors also reference SDI as an example of ambitious U.S. programs that fail to move past the prototype or testing phase, let alone the successful deployment of a system.[136] The Russian narrative thus emphasizes that Russia maintains an assured nuclear second-strike capability—and is actually more successful in its technological advancements in space, according to many articles that we surveyed—but must act on behalf of weaker countries who lack Russia's technological prowess.[137]

In the articles from *Military Thought*, the authors are relatively more willing to acknowledge that the United States (at the time) possessed the world's most-powerful space capabilities. The authors focus less on defending Russian identity as a great power in space and more on emphasizing the previous, dangerous results of arms races throughout history.[138] The articles in *Military Thought* that we reviewed also draw a contrast between the strategic documents of the United States and those of Russia to show that Russia conceptually lagged the United States at the time. The authors articulate the perception that while U.S. national space policy demonstrates that the United States understands and has capitalized on the role of air and space capabilities in future war, Russia's doctrine does not even consider space as a warfighting domain. One author notes, for instance, "Unlike the Americans, for some reason we do not associate national security with space activities, as it can be seen in the 'National Security Concept,' ratified by the Presidential Decree No. 24 from January 10, 2000, and which does not have a single word about space."[139]

Operation Burnt Frost (2008)

Event Overview

On February 20, 2008, a Standard Missile-3 (SM-3) that was fired from the USS *Lake Erie* targeted USA-193, a nonfunctioning satellite tumbling toward earth. USA-193 malfunctioned at launch on December 17, 2007, and its 1,000 pounds of hazardous propellant were expected to survive the satellite's reentry. Although the USS *Lake Erie* received the primary tasking, two additional SM-3 missiles on two additional ships—the USS *Decatur* and USS *Russell*—were tasked to provide a backup capability. The satellite and the tank of hazardous propellant were

---

[135] Vladimir Sidorov, "Inostrannyye Voyennyye Novosti" ["Foreign Military News"], *Krasnaia Zvezda* [*Red Star*], No. 57, 2005; and Vladimir Kozin, "SSHA Rassmatrivayut Kosmos Kak Budushcheye 'Pole Boya'" ["US Sees Space as a Future 'Battlefield'"], *Krasnaia Zvezda* [*Red Star*], No. 29, 2018.

[136] Viktor Ruchkin, "Bez 'Zvezdnykh Voin'" ["No Star Wars"], *Krasnaia Zvezda* [*Red Star*], No. 175, 2009.

[137] Sidorov, 2005; Ruchkin, 2009.

[138] Chel'tsov, 2007; and "Uroki i Vyvody iz Voyny v Irake" ["Lessons and Conclusions from the War in Iraq"], *Voennaia Mysl'* [*Military Thought*], No. 7, 2003.

[139] "Uroki i Vyvody iz Voyny v Irake" ["Lessons and Conclusions from the War in Iraq"], 2003.

successfully destroyed, with no permanent space debris.[140] This came one year after China's first direct-ascent kinetic-kill vehicle ASAT test in January 2007, which generated significant space debris that remains in orbit today.

China

Operation Burnt Frost [燃霜行动] is generally perceived in China as yet another example of U.S. militarization of space and, specifically, as demonstrating a kinetic counterspace capability. It is not widely discussed in PLA texts, is extremely rarely referenced explicitly as "Operation Burnt Frost," and is not directly referenced in any of the key PLA books or broader PLA articles we reviewed.[141] One 2015 article by a PLA Second Artillery Command College researcher mentions the event in passing, referencing it as a shipborne test of a missile intercepting a satellite [舰载导弹拦截卫星试验] and stating it "proved that [the United States] has the ability to implement kinetic energy interception of satellites from any point in the oceans of the world."[142] The operation also appears to have reinforced Beijing's perspective that U.S. missile defense efforts hold an offensive capability, if not intent.[143] Although the SM-3 was not designed as a counterspace capability and required modification for the 2008 operation, it was described in

---

[140] Nicholas L. Johnson, "Operation Burnt Frost: A View from Inside," *Space Policy*, Vol. 56, May 2021; and Missile Defense Agency, U.S. Department of Defense, "Aegis Ballistic Missile Defense: One-Time Mission: Operation Burnt Frost," undated.

[141] It is referenced in some PLA articles, but, unfortunately, most of the articles that mentioned it were not available to us. See Li Daguang [李大光], "Looking at the Development of Anti-Satellite Weapons from the U.S. 'Missile-to-Satellite'" ["由美国'导弹打卫星'看其反卫星武器发展"], *Defense Industry Conversion in China* [中国军转民], July 2010; Huo Mu [火木], "What Is the True Intention of the United States to Destroy Runaway Satellites with Sea-Based Missiles?" ["美国用海基导弹摧毁失控卫星真实意图何在?"], *Modern Navy* [当代海军], April 2008, pp. 22–23; Li Daguang [李大光], "Look at the Development of Anti-Satellite Weapons from the United States' 'Missile-to-Satellite'" ["由美国'导弹打卫星'看其反卫星武器发展"], *Technology Foundation of National Defence* [国防技术基础], July 2008; Wang Shusheng [王树生] and Yang Xuefeng [杨学锋], "'Patriot' Boarded the Ship, and the U.S. Navy's Anti-Missile Capability Was Enhanced" ["'爱国者'上舰，美海军反导能力提升"], *Shipborne Weapons* [舰载武器], September 2008, pp. 10–11; Li Shuang [李爽], *A Study on the Code of Conduct in Outer Space from the Perspective of Governance Theory* [治理理论视阈下的外空活动行为准则研究], master's thesis, Changsha, China: PLA National University of Defense Technology [国防科学技术大学], 2014; Wei Chenxi [魏晨曦], "Development Trend of Space Operations from Schriever Wargame" ["从'施里弗'系列演习看未来太空作战的发展"], *Space International* [国际太空], June 2016; Zhang Xuesong [张雪松], "The United States Builds a Space Force That Has Reached the Point of No Return" ["美国组建天军如箭在弦"], *Space Exploration* [太空探索], September 2018; Zhang Yan [张岩], "India Tests Anti-Satellite Weapon" ["印度成功试验反卫星武器"], *Aerospace China* [中国航天], May 2019; and Zhang Ming [张茗], "U.S. National Security Space Strategy Pivot and Its Implications for China" ["美国太空安全战略转向及其对中国的影响"], *Journal of Social Sciences* [社会科学], September 2020.

[142] Guo Jun [郭俊], "Construction of U.S. Space Deterrence Force Revealed in 'Schriever' Exercises" ["从'施里弗'演习看美军太空威慑力量构建"], *National Defense Technology* [国防科技], Vol. 36, No. 1, February 2015, pp. 68–70, 89.

[143] For more on this, see Gregory Kulacki and Jeffrey G. Lewis, "Understanding China's Antisatellite Test," *Nonproliferation Review*, Vol. 15, No. 2, 2008.

2021 as a "space weapon" by the director of the U.S. Missile Defense Agency.[144] In terms of China's public response, the MFA criticized the 2008 event and said, "China is continuously following closely the possible harm caused by the U.S. action to outer space security and relevant countries."[145] Of note, there was no observed Chinese discussion of Operation Burnt Frost as a response to China's own 2007 direct-ascent ASAT test.

Russia

The articles in the *Red Star* data set characterize Operation Burnt Frost as a specific example of U.S. aggressiveness and the U.S. desire to militarize space for its own purposes. Yet again, though, Russian authors in *Red Star* try to strike a delicate rhetorical balance. First, the articles challenge the U.S. claim to space superiority, arguing China and Russia were the first to successfully strike a satellite with a missile.[146] The authors thus contend that the United States is playing catch-up with this ASAT strike. Some authors further contend that the United States knew that other states would interpret this mission as the demonstration of an ASAT capability, despite public rhetoric from U.S. officials characterizing the strike as an act of responsibility.[147]

Consistent with Russian perceptions of the 2006 U.S. National Space Policy and SDI, *Red Star* authors again leveraged critiques from American sources to buttress their arguments. For instance, a 2008 article incorrectly quotes Ivan Oelrich, Vice President for the Strategic Security Project in the Federation of American Scientists, as saying, "[the] destruction of this satellite may be just an excuse for testing anti-satellite weapons."[148]

Finally, descriptions of Operation Burnt Frost are particularly noteworthy to the degree that interest in the event has faded with time. The articles that mention Operation Burnt Frost are clustered around the time when it happened.[149] Unlike mentions of SDI, for example, which are present throughout the data set, this event was replaced by the narratives on threats posed by other U.S. activities, specifically the X-37B, Space Fence, and, later, the USSF.

---

[144] Jon A. Hill and Michelle C. Atkinson, "Department of Defense Press Briefing on the President's Fiscal Year 2022 Defense Budget for the Missile Defense Agency," press conference transcript, U.S. Department of Defense, May 28, 2021.

[145] Keith Bradsher, "China Criticizes U.S. Missile Strike," *New York Times*, February 22, 2008.

[146] Sidorov, 2009.

[147] Viktor Ruchkin, "Pentagon Nachinayet 'Zvezdnyye Voyny'" ["Pentagon Starts the 'Star Wars'"], *Krasnaia Zvezda* [*Red Star*], No. 27, 2008.

[148] Vladimir Kozin, "Aziatskiy 'Dovesok' k YEVROPRO" ["Asian 'Add-on' to European Missile Defense"], *Krasnaia Zvezda* [*Red Star*], No. 69, 2008. The correct quote from Oelrich is the following: "We should be working toward a treaty to ban anti-satellite tests, not looking for dodgy reasons to conduct them" (Ivan Oelrich, "Transcript: Spy Satellite Shootdown," *Washington Post*, February 20, 2008).

[149] Note that we found no mentions in *Military Thought* after conducting several searches using search terms in both languages (*USA-193*; *SM-3*; and more-generic terms, such as *shoot* and *satellite*). The absence of articles on this topic might signify that this event was of less interest for the more research- and analysis-oriented community in Russia.

President Obama's National Space Policy (2010) and National Security Space Strategy (2011)

Event Overview

The Obama administration's National Space Policy emphasized inclusiveness and the need for all nations to cooperate in the responsible use of space. The policy focused on six goals of the U.S. space program: energizing a competitive domestic industry, expanding international cooperation, strengthening stability in space, increasing assurance and resilience of essential functions enabled by space-based assets, pursuing innovation in robotics, and improving space-based earth and solar observation.[150]

China

Although the Obama administration's space policies were generally viewed in Beijing as an adjustment of U.S. tone and approach, the policies were perceived as not altering the fundamental U.S. intent: the pursuit of space dominance.[151] For example, one writer in 2010 argued that while the Obama administration's space policy had changed some goals set during the Bush administration, such as increasing international cooperation in space, it did not alter the U.S. goal of achieving space hegemony.[152] Another author pointed out that while, on the one hand, the Obama administration had shifted away from the Bush administration's unilateral approach in favor of multilateral cooperation, on the other hand, the 2011 U.S. National Security Space Strategy [2011 美国国家安全太空战略] espoused strengthening U.S. leadership in space and maintaining the U.S. strategic advantage.[153]

Further emphasizing the point, a 2011 article argued that the 2010 U.S. National Space Policy

> emphasized the expansion of international cooperation, the enhancement of space security, and the development of commercial space. Compared with the space policy promulgated by the Bush administration in 2006, the new version of the "National Space Policy" adjusted the relevant content with unilateralism, and the tone and wording were relatively calm.[154]

---

[150] The White House, *National Space Policy of the United States of America*, Washington, D.C., June 28, 2010.

[151] None of the key policy documents of this era were prominently featured in the primary Chinese texts that we surveyed, but they are covered elsewhere and are generally summarized accurately.

[152] Cheng Qun [程群], "On the Adjustment of the Obama Administration's Space Policy" ["浅析奥巴马政府太空政策的调整"], *Contemporary International Relations* [现代国际关系], June 2010, p. 33; and DoD and Office of the Director of National Intelligence, *National Security Space Strategy: Unclassified Summary*, Washington, D.C., January 2011.

[153] Gaoyang and Ke, 2014, pp. 1–5, 45.

[154] Qu Jing [曲晶] and Ge Shanshan [葛姗姗], "Development of Foreign Launch Vehicles in 2010" ["2010 年世界航天运载器发展回顾"], *Missiles and Space Vehicles* [导弹与航天运载技术], February 2011.

Nevertheless, the Obama administration's push for international norms and international cooperation, in this view, is simply another way to "establish a new international space order led by the United States and further consolidate its leading position in space."[155]

A 2012 article addressed both the 2010 National Space Policy and the 2011 National Security Space Strategy, stating,

> [The Chinese] recognize the U.S. need for the stability of the space environment, abandon George W. Bush's "unilateral" policies, and hope to formulate an international space code of conduct by enhancing the sharing of space situational awareness and information, and extensively carry out international cooperation and other means to enhance the stability of the space environment.[156]

A similar 2013 article relayed that the 2011 strategy said, "The United States will work with allies and partners to ensure safe, sustainable, and stable access to and use of space in order to achieve the United States' space security goals and maintain its leading position in space."[157]

The 2010 National Space Policy was directly cited only three times in our Chinese literature sample, perhaps because it was perceived as less threatening and thus less noteworthy. As with SDI, this might be a further indication of confirmation bias because PLA scholars actively search for a U.S. threat to the PRC. Authors simply summarized the 2010 National Space Policy and did not use the hyperbolic language that other authors employed when discussing the 2006 National Space Policy or SDI. The authors took interest in the U.S. description of the current strategic environment in space as being "crowded," "contested," and "competitive."[158] All three articles were clear that the United States was basing its future space strategy on the "3C." One article did note that the strategic principles laid out in the 2010 policy appeared to conform to international norms but, upon closer inspection, especially regarding the claim that the United States would take various actions to deter interference and attacks between space-based systems, revealed that the United States still sought space advantage [太空优势].[159]

The central argument of the articles, which deal with the overall development of U.S. space strategy, is that the United States is striving for space hegemony. This might indicate that the authors saw the 2010 National Space Policy as less provocative than either SDI or the 2006 U.S. National Space Policy. Chinese authors in our sample did appear to overlook the 2010 policy's guidance to the Secretary of Defense on counterspace actions, including to "develop capabilities, plans, and options to deter, defend against, and, if necessary, defeat efforts to interfere with or

---

[155] Qu and Ge, 2011.

[156] Xia Yu [夏禹], "Focusing on Future Space Operations: NATO Participates in Schriever 2012 Exercise" ["聚焦未来空间作战:北约参加施里佛 2012 演习"], *Satellite Application* [卫星应用], April 2012.

[157] Long Xuedan [龙雪丹] and Qu Jing [曲晶], "Review of World Launch Vehicles in 2012" ["2012 年世界航天运载器发展回顾"], *Missiles and Space Vehicles*, February 2013.

[158] Gaoyang, 2018, pp. 116–130, 135.

[159] Fan and Gong, 2016a, pp. 40–44.

attack U.S. or allied space systems [and] maintain the capabilities to execute the space support, force enhancement, space control, and force application missions."[160] That said, these PLA articles still cited the 2011 U.S. National Security Space Strategy as another step toward U.S. control of space. All of this suggests that the Chinese perceived both Obama administration documents as less inflammatory than the Bush-era documents but as still not deviating from the greater trend of the United States attempting to increase its control of space.

Russia

The tone of the articles in the data set that address the Obama administration's space policy, especially the tone of the articles published earlier in the Obama presidency, is markedly different from the tone used when discussing the other events selected for this project. U.S. space policies are described positively, and Obama is portrayed as a beacon of hope for a more peaceful shift of space efforts.[161] After President Bush's bellicose rhetoric of space superiority, President Obama's approach was perceived by Russian authors as a chance to finally normalize the relations between great space powers. President Obama was even perceived as being capable of discontinuing the trends in the United States toward space militarization and weaponization. According to a 2010 article, President Obama explicitly stated he was going to reverse the space policy of the Bush administration. The author also quoted President Obama as talking about the need to protect military and civilian satellites without distinguishing between the two.[162]

Moreover, Russian authors emphasized the Obama administration's acknowledgment of the importance of finding a diplomatic solution to ensure that space remained peaceful. One author explicitly drew a contrast between President Obama and his predecessor, noting that President Obama spoke in favor of international regulation of space behavior. The positive perception of U.S. words is moderated, though, by concern about the challenges of moving in a less aggressive direction. The author advised caution, arguing the existing space doctrine directly prohibited engagement in any efforts aimed at demilitarizing space, which was perceived as creating a potentially insurmountable obstacle for President Obama's attempted policy change.[163]

The initial Russian perception of optimism was also disrupted by the critiques of Obama administration policy by U.S. sources. Given this criticism, Russian authors perceived that the United States would not entirely abandon the desire to dominate in space. Moreover, some authors claimed the United States continued to work on space defensive and offensive weapons, despite changing rhetoric.[164] Russian authors thus suspiciously watched for signs in subsequent

---

[160] The White House, 2010, p. 14.

[161] Sergey Medvedev, "Kosmicheskiy Proryv Obamy" ["Obama's Space Breakthrough"], *Krasnaia Zvezda* [*Red Star*], No. 66, 2010.

[162] "Voyennaya Platforma Obamy" ["Obama's Military Platform"], *Krasnaia Zvezda* [*Red Star*], No. 207, 2008.

[163] "Voyennaya Platforma Obamy" ["Obama's Military Platform"], 2008.

[164] Medvedev, 2010.

statements and strategic documents, arguing some statements pointed to a potentially even more threatening U.S. approach to space than the policies of the Bush administration.[165] For example, President Obama was quoted as allegedly saying, "In the coming years, the United States intends not only to continue moving along the beaten path but to make a leap into the future."[166] In the perception of this Russian author, this statement was an ominous signal the administration was abandoning the "friendly" approach of the first years of the Obama presidency for the more typical U.S. rhetoric of competition in space and treatment of other space powers as adversaries.

General Shelton's Remarks Regarding the Geosynchronous Space Situational Awareness Program (2014)

Event Overview

General William Shelton publicly discussed GSSAP for the first time on March 3, 2014. The USAF stated that the purpose of GSSAP satellites was to assist in the tracking of manmade objects in geosynchronous orbit, where many of the U.S. military's strategic communications and early warning satellites reside.[167] General Shelton described GSSAP as a "neighborhood watch" for satellites, which the USAF described as uniquely contributing to "timely and accurate orbital predictions, enhancing our knowledge of the geosynchronous orbit environment, and further enabling space flight safety to include satellite collision avoidance."[168] The satellites would also be capable of maneuvering to inspect objects in this orbit.[169] GSSAP, and General Shelton's public remarks that it existed, came after China had been deploying maneuverable satellites, potentially representing a co-orbital ASAT capability, since the launch of Shijian-12 in June 2010.

China

GSSAP [地球同步轨道空间态势感知计划] is generally viewed as a key part of the U.S. military's space operations and, thus, another piece of the U.S. militarization of space. The 2021 PLASSF SSD book frames GSSAP as one of several U.S. space-based situational awareness systems and notes that such satellites are "important nodes of the entire space information system

---

[165] Vladimir Kozin, "Pol'sha, Rumyniya... Kto Sleduyushchiy?" ["Poland, Romania... Who's Next?"], *Krasnaia Zvezda* [*Red Star*], No. 21, 2010.

[166] Viktor Ruchkin, "V Kosmose Tainstvennyy Chelnok" ["The Mysterious Shuttle in Space"], *Krasnaia Zvezda* [*Red Star*], No. 72, 2010.

[167] Stephen Clark, "Air Force General Reveals New Space Surveillance Program," Space.com, March 3, 2014.

[168] Clark, 2014.

[169] Clark, 2014.

and are of great strategic significance."[170] It argues GSSAP satellites can "re-enter and drift up and down in synchronous orbits, and carry out large-scale orbital maneuvers to approach the targets for reconnaissance and surveillance [and can] provide the U.S. military space surveillance and combat operations support."[171] It also adds,

> In short, as congestion and competition in the geostationary orbit intensify, the U.S. military believes that GSSAP can prevent aggression against GEO [geosynchronous orbit] and provide timely and accurate intelligence data to achieve space defense missions to protect U.S. and allied spacecraft targets.[172]

One 2018 article we reviewed similarly frames GSSAP as part of the continued U.S. militarization of space and notes that the GSSAP satellites were thought to monitor space targets in geosynchronous orbit and be able to conduct rendezvous and proximity operations [交会接近行动].[173] Another 2018 book on global aerospace developments argues,

> [T]he satellites have high-precision orbital maneuvering capabilities. They can maneuver near the earth's synchronous orbit, perform close-up and detailed imaging reconnaissance of targets, and even intercept electronic signals to provide target technical reconnaissance and action intention judgment for U.S. military space operations.[174]

Russia

Red Star closely follows the development of assets in the U.S. military satellite network. Developments are described both in brief news articles and in more-elaborate analytical articles that assess certain events for insights. The brief news articles often directly follow the launches of the U.S. military satellites and emphasize that a satellite's mission either is secret or has an active military purpose but that the satellite is disguised as, for instance, a satellite for monitoring space debris. The active military purposes ascribed by Russian authors include the ability of GSSAP satellites to observe other satellites for suspicious activities, the use of these satellites to create a foundation for deploying strike capabilities in space in the future, and their use as a

---

[170] Feng and Dong, 2021, pp. 212–216. For another article, see "U.S. Completes Geosynchronous Space Situational Awareness Program Ground System Upgrade" ["美完成地球同步轨道太空态势感知计划地面系统升级"], *Aerospace Electronic Warfare* [航天电子对抗], Vol. 36, No. 2, 2020, p. 13.

[171] Feng and Dong, 2021.

[172] Feng and Dong, 2021.

[173] Gaoyang, 2018, pp. 116–130, 135.

[174] *Report on Scientific Developments in the Aerospace Field: 2017* [航天领域科技 发展报告], Beijing: National Defense Industry Press [国防工业出版社], April 2018, pp. 60–61.

latent threat that can easily be transformed into a weapon.[175] *Red Star* authors also perceive such satellites as decreasing the transparency of space activities.[176]

Some articles connect the launch of satellites in the GSSAP program to the creation of the USSF. The articles stress that the United States is increasingly monitoring the activities of other spacefaring nations. The tone and substance of the articles also seem to reflect the perception that GSSAP satellites—which, as mentioned above, are maneuverable and located in geosynchronous orbit—create a significantly threatening capability. Authors argue that these features allow the satellites to fly closely (and covertly) to other military and civilian space objects to inspect them and collect information.

Of particular note with respect to GSSAP is how similar the Chinese and Russian perceptions are regarding this event. Beijing and Moscow express similar concerns about the putative ulterior motives enabled by GSSAP satellites. Both believe the architecture of satellites lays the groundwork upon which a more robust future capability can be built, a capability that would allow the United States to achieve a strategic effect in a future conflict: domination of the information environment.

As mentioned earlier, Figures 2.1–2.4 provide a quantitative illustration of Chinese and Russian writing over time regarding selected U.S. space activities.

---

[175] See, for example, "Inostrannaya Voyennaya Khronika" ["Foreign Military Chronicle"], *Krasnaia Zvezda* [*Red Star*], No. 86, 2015; and Anna Polyakova, "Imperativy 'Tsarya Gory'" ["The Imperatives of the 'King of the Hill'"], *Krasnaia Zvezda* [*Red Star*], No. 167, 2015. The articles note that other countries that also conduct space monitoring report active movement of GSSAP satellites, including in proximity to the other spacecraft.

[176] "V Inostrannykh Armiyakh" ["In the Foreign Armed Forces"], *Krasnaia Zvezda* [*Red Star*], No. 64, 2021.

Figure 2.1. Chinese Interest in Selected U.S. Space Activities over Time

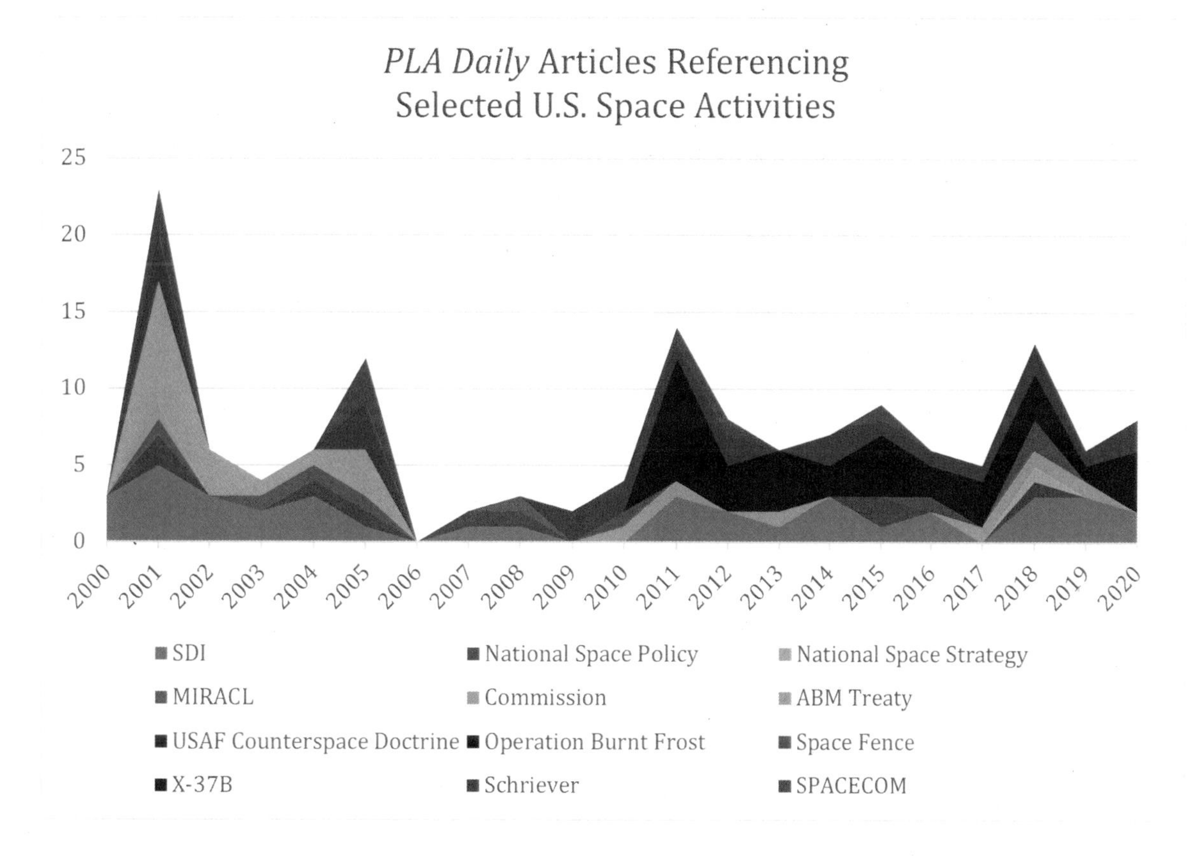

SOURCE: Authors' count of data from China National Knowledge Infrastructure (CNKI) database.
NOTES: SPACECOM = U.S. Space Command. Data are for *PLA Daily* articles that reference the specific U.S. space activity (various translations are provided throughout this chapter).

Figure 2.2. Russian Interest in Selected U.S. Space Activities over Time

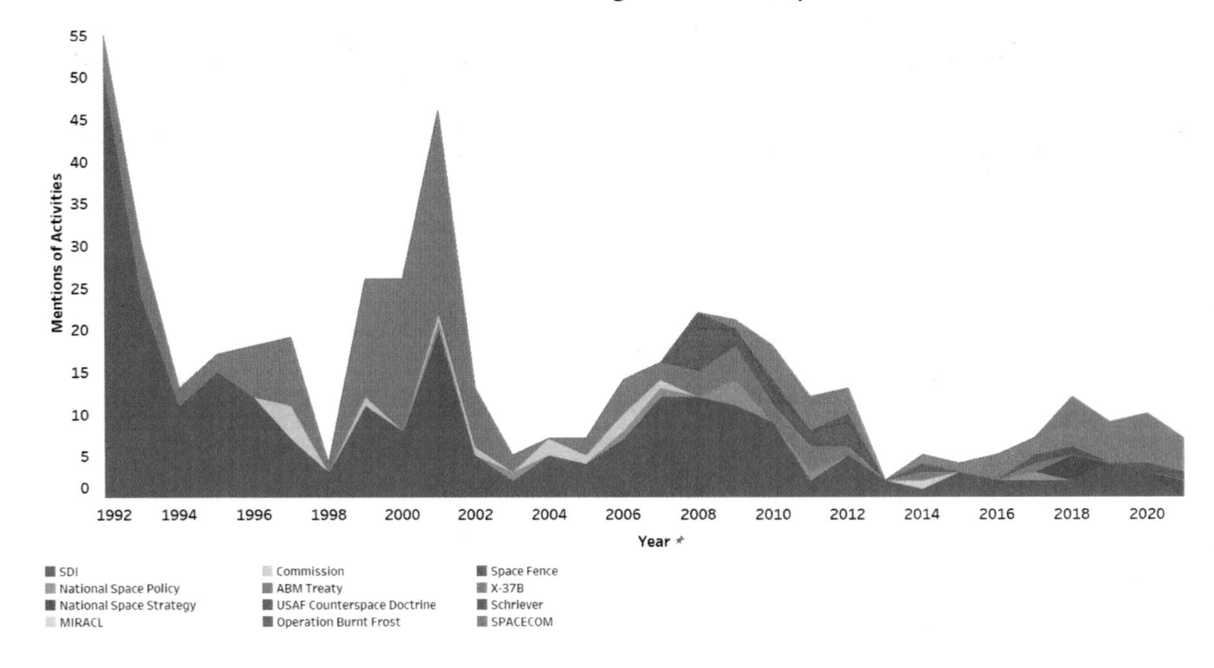

SOURCE: Authors' count of data from East View Information Services Online database.
NOTE: SPACECOM = U.S. Space Command.

Figure 2.3. Chinese Interest in U.S. Administrations and Space over Time

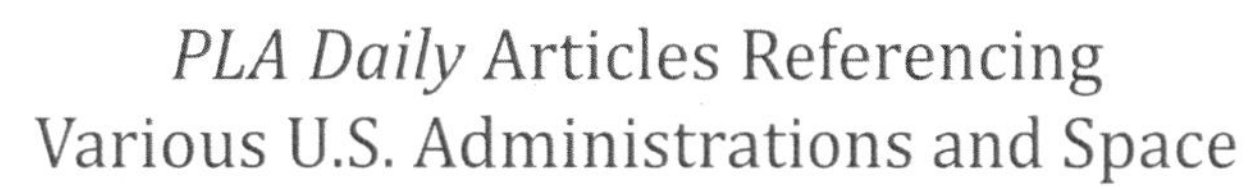

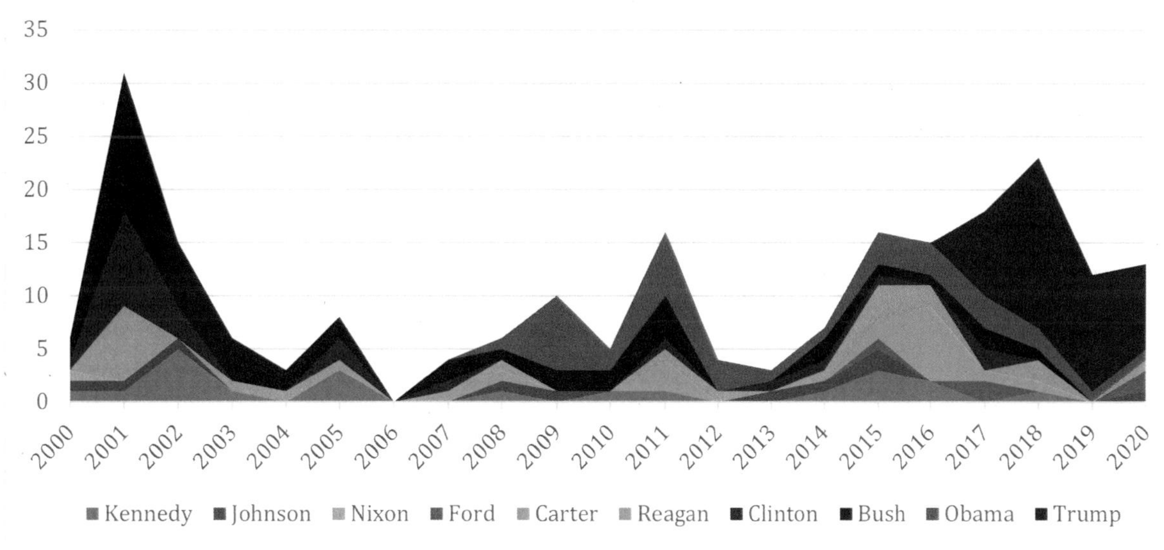

SOURCE: Authors' count of data from CNKI database.
NOTES: Data are for *PLA Daily* articles that reference "space" [太空 or 外空] and the specific U.S. administrations. This figure combines the George H. W. Bush and George W. Bush administrations, since PRC writings are not always consistent in identifying the specific Bush administration.

Figure 2.4. Russian Interest in U.S. Administrations and Space over Time

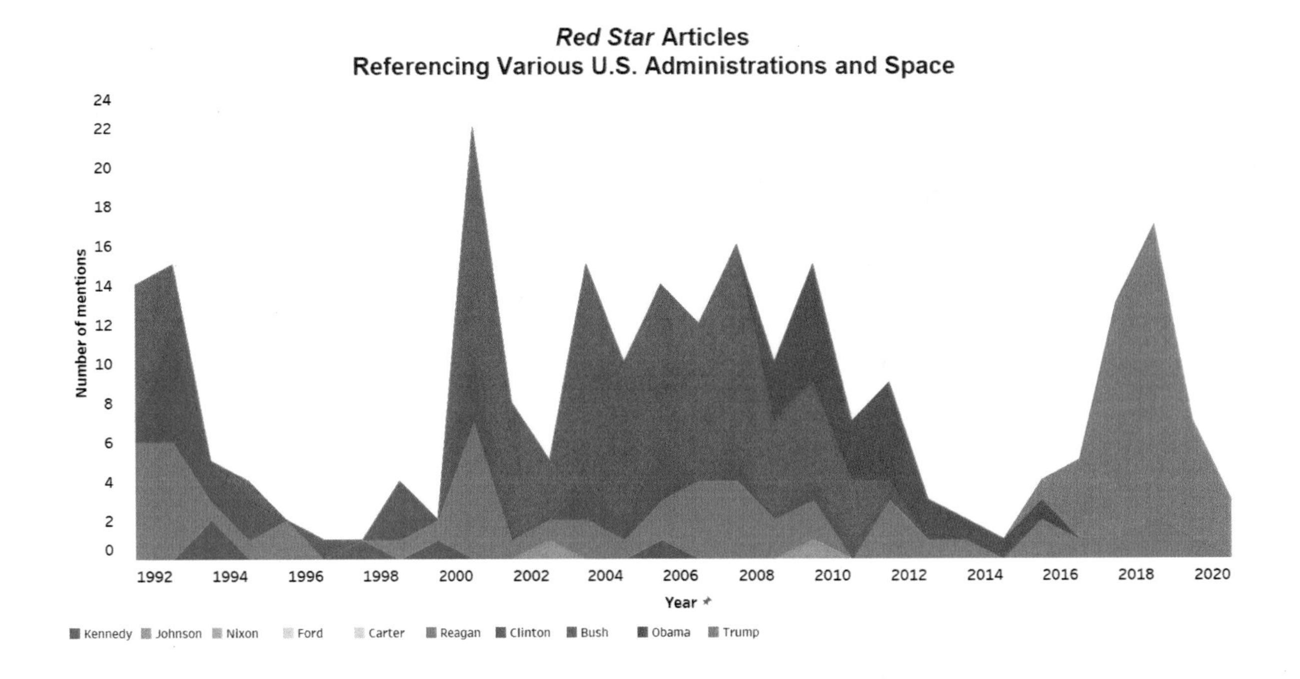

SOURCE: Authors' count of data from East View Information Services Online database.
NOTE: This figure combines the George H. W. Bush and George W. Bush administrations, since Russian writings are not always consistent in identifying the specific Bush administration.

Chapter 3. Implications

Having completed a survey of a meaningful (but necessarily circumscribed) portion of the native-language Chinese and Russian literature regarding U.S. activities and policies with respect to space, we conclude this report with our assessment of the implications. We leverage our improved understanding of how Chinese and Russian perceptions about U.S. activities in space have evolved over time and of the responses taken to address or counter U.S. actions to analyze what the data sets suggest regarding the following questions:

- Do Chinese and Russian concerns appear to be genuine?
- Do China and Russia appear to have artificially exaggerated their concerns to delegitimize U.S. space activities?
- Do Chinese and Russian actions in space appear to be in response, at least in part, to U.S. space activities?

Evolving Perspectives, Consistent Perception of Threat

The Chinese and Russian native-language primary sources reviewed for this project reflect a sustained perception that U.S. military activities related to the space domain are threatening and reflect hostile U.S. intent. This finding is consistent with the political science literature, specifically the action-reaction model, which hypothesizes that states react to the arming of other states, fueled by uncertainties about intentions and capability.[177] It is also consistent with the contention that states tend to overestimate the capabilities of their adversaries.[178] These two insights help us understand why China and Russia appear to perceive U.S. military activities in the space domain as threatening and that this perception has been the consistent narrative reflected in native-language primary sources since SDI.

Yet U.S. actions also seem to be interpreted through the lens of the state of the relationship at a given point in time. Perhaps unsurprisingly, when the political relationship between Washington and Beijing or Moscow is more confrontational and fractious, the same actions are perceived to be more threatening, and when bilateral relations are (relatively) more friendly, similar actions are not viewed as quite as threatening. Moreover, while the perception of threat has been consistent over time, there has been a deepening of the sense of hostility since 2008, following the issuance of the 2006 National Space Policy and Operation Burnt Frost. Chinese

[177] Barry Buzan and Eric Herring, *The Arms Dynamic in World Politics*, Boulder, Colo.: Lynne Rienner Publishers, 1998; and George W. Rathjens, "The Dynamics of the Arms Race," *Scientific American*, Vol. 220, No. 4, April 1969.

[178] Jonathan Renshon, "Assessing Capabilities in International Politics: Biased Overestimation and the Case of the Imaginary 'Missile Gap,'" *Journal of Strategic Studies*, Vol. 32, No. 1, February 2009.

and Russian analyses reviewed for this report generally noted a more positive, cooperative approach toward space by the Clinton and Obama administrations but still focused on their perceived overarching continuity of hostile and aggressive U.S. military space policy. This suggests that existing negative perceptions held in Beijing and Moscow are relatively easily reinforced by those U.S. actions perceived as hostile, while U.S. actions perceived as less hostile do not appear to have a similarly robust effect, producing a seemingly minimal improvement in Chinese and Russian perceptions. This behavior might reflect the natural, human tendency toward confirmation bias. Yet it is important for U.S. decisionmakers to understand that this is occurring.

In the case of China, Beijing has always been concerned about U.S. space policy, but its assessment of U.S. capabilities appears to have worsened. As early as 1995, China's defense white paper stated that "the major nuclear powers, with the world's most sophisticated and largest quantity of nuclear weapons in hand, have neither abandoned their policy of nuclear deterrence nor stopped the development of nuclear weapons and outer space weapons including guided missile defense systems."[179] Such criticism continued through the 2000s; the 2002 defense white paper noted, "At present, outer space is faced with the danger of weaponization."[180] The 2015 defense white paper suggested that this long-standing risk was finally becoming a reality: The "countries concerned [i.e., the United States] are developing their space forces and instruments, and the first signs of weaponization of outer space have appeared."[181] China appears to pay more attention to U.S. space actions and programs than U.S. space policies and strategy documents, though the latter are often interpreted as self-fulfilling prophecies that rationalize underlying U.S. hostile intent and provide either justification for or a facade over U.S. space actions.

Russia's perception of the increasingly hostile nature of the threat posed by U.S. activities in space can be discerned in the evolution of the language used in Russian military doctrine. The 2000 *Military Doctrine of the Russian Federation* simply mentions "control of outer space" and "missile defense," among other threats, as "actions to undermine global and regional stability," while the 2010 *Military Doctrine of the Russian Federation* identifies the "increased military use of airspace and outer space" as a main external danger.[182]

The 2014 *Military Doctrine of the Russian Federation* lists the "establishment and deployment of strategic missile defense systems undermining global stability and violating the

---

[179] PRC Information Office of the State Council, "China: Arms Control and Disarmament," November 1995.

[180] PRC Information Office of the State Council, 2002.

[181] PRC Information Office of the State Council, "China's Military Strategy," Xinhua, May 2015.

[182] Security Council of the Russian Federation, *Voyennaya Doktrina Rossiyskoy Federatsii* [*The Military Doctrine of the Russian Federation*], No. 706, Moscow, April 21, 2000; and Security Council of the Russian Federation, *Voyennaya Doktrina Rossiyskoy Federatsii* [*The Military Doctrine of the Russian Federation*], No. 146, Moscow, February 5, 2010.

established balance of forces related to nuclear missiles, implementation of the global strike concept, intention to place weapons in outer space, as well as deployment of strategic non-nuclear systems of high-precision weapons" as one of the "main external military risks" to the Russian Federation.[183] The doctrine further states that one of the main tasks for the Russian Federation to deter and prevent future conflict is "to resist attempts by some states or group of states to achieve military superiority through the deployment of strategic missile defense systems, the placement of weapons in outer space or the deployment of strategic non-nuclear high-precision weapon systems."[184] These statements appear to reflect a relative increase in Russia's perception of the hostility of the external threat environment.

The key overarching factor shaping Chinese and Russian perceptions of the U.S. threat appears to be the longer-term military *potential* of various U.S. activities regarding space rather than the immediate impact of a specific activity or policy at that point in time. China and Russia appear to be hedging their bets that technology may eventually catch up with perceived U.S. intentions to militarize space. For example, the Russian sources surveyed appear to have readily discarded the technical feasibility of SDI at the time of its development but nonetheless expressed concern about the *implication* of such an initiative—i.e., that it suggests that the United States seeks to militarize space.

Two key concerns appear to drive this Chinese and Russian threat perception: nuclear deterrence and counterspace. First, both countries often view U.S. military space-related activities as threatening to their respective nuclear deterrents. This appears to have its origins in SDI, which both countries viewed correctly as fundamentally about using the space domain to support and supplement U.S. nuclear capabilities. As nuclear competition with Beijing and Moscow heats up in the coming years, it will be important for U.S. policymakers to understand that U.S. military activities in space—regardless of U.S. intent—may be linked to U.S. nuclear capabilities and strategy. Second, both China and Russia are very concerned about U.S. counterspace intent and capabilities, such as Operation Burnt Frost. This view also relates to concern over the ability of U.S. satellites to fly closely (and covertly) to other space objects to inspect them and collect information, as well as concern over broader U.S. space domain awareness intent and capabilities, such as GSSAP. One additional point is that both countries also view U.S. military actions as critical enablers to U.S. conventional military operations, such as providing space-based intelligence, surveillance, and reconnaissance for long-range precision-guided munitions in the Gulf War.[185]

While some might argue these perceptions are not genuine—that they are merely elaborate propaganda to hype the existence of a U.S. threat—the immense breadth of source materials and their broad consistency over time strongly suggest that these concerns, as documented in very

---

[183] President of the Russian Federation, 2014.

[184] President of the Russian Federation, 2014.

[185] For some Chinese examples, see Chang, 2005; and Jiang, 2013, pp. 10, 252.

niche and often internally focused Chinese and Russian primary sources, reflect the consensus view within the Chinese and Russian governments.[186] This is further borne out in some of the action-reaction responses by China and Russia to U.S. space initiatives over time, discussed more below. Regardless of whether U.S. analysts agree with these foreign perceptions, they should accept them as genuine and account for them as they consider future activities.

Threading the Rhetorical Needle: Threatening Versus Threatened

While the perception of threat certainly appears to be genuine (or at least to be the internal consensus), it is more difficult to assess whether Chinese and Russian public reactions are intentionally inflated. One example that suggests there is some measure of intentionally inflated rhetoric is the evolving narrative offered by a single Russian analyst regarding Operation Burnt Frost. In an early article, the author stressed the relative rarity of a malfunctioning satellite and used this observation to conclude that the situation might signify the incompetence of the U.S. space program, with dangerous consequences for the world community. Although this initial article detailed at length the potential threats from a falling satellite, after the United States destroyed the satellite with a retrofitted missile, this context disappeared from a later article in *Red Star* by the same author, who argued the incident was an excuse to test space strike weapons.[187] In the later article, the author asserted the United States intentionally exaggerated the possible dangers to appear like a "savior" of humanity [*spasitel' mira*] rather than a potential aggressor.[188]

As reflected in our data set, both China and Russia appear to attempt to strike their own different and delicate rhetorical balancing acts through nuanced arguments that characterize U.S. actions as threatening while characterizing their own, similar actions as nonthreatening. For example, although China has long criticized U.S. space militarization, Beijing's MFA defended China's 2007 ASAT test, 12 days after the fact, saying it was "not directed at any country and does not constitute a threat to any country."[189] For its part, Russia describes U.S. actions in space as increasingly hostile and aggressive, focusing on specific actions, such as the MIRACL test and Operation Burnt Frost, to demonstrate that specific U.S. policies, despite U.S. rhetoric to the contrary, are aggressive.

Russia's balancing act centers on denouncing the United States as aggressive in space while claiming to be an equal space power that does not fear the United States. In the literature we

---

[186] While one might argue that the internal consensus view thus should be considered a "genuine" perception, it is possible, though difficult to determine, that the sources reviewed here and broader Chinese and Russian analyst communities adjust their views to match others' (e.g., senior leaders') preconceived beliefs. Regardless, what matters is the fact that the Chinese and Russian systems perpetuate this internal perception.

[187] Kuzar', 2008a; Kuzar', 2008b.

[188] Kuzar', 2008a.

[189] Joseph Khan, "China Confirms Test of Anti-Satellite Weapon," *New York Times*, January 23, 2007.

reviewed, Russian authors appeared to be highly sensitive to the potential implication that Russia is not an equal of the United States in the space realm. This is perhaps unsurprising given that space plays a significant role in the Russian national psyche, representing heroism, pride, status, and strength.[190] Thus, characterizations of a threat posed by the United States are moderated by declarations that *other* states are menaced by U.S. actions, not Russia per se. In the sources that we reviewed, Russia thus perceives its actions as unthreatening and perceives itself as coming to the aid of the international community, using its spacefaring technical prowess to counter U.S. moves to militarize space.

Even while sounding the alarm, though, Russia continues to emphasize cooperation with the United States in the peaceful exploration of space. Because the two countries have cooperated on the International Space Station since 1993, Moscow cannot frame *all* U.S. space actions as threatening. As recently as 2017, for example, Russian President Vladimir Putin stated, "we are planning to do this [future space projects at the new Vostochny Cosmodrome in Russia's Far East] with the United States, unless 'wiseacres' from various American structures interfere. . . ."[191] These continued calls for scientific cooperation, amid the perception of a rising U.S. military threat in space, suggest that Russia views space as a vehicle for demonstrating its international leadership bona fides and its great-power status.[192]

The tension for Moscow is between advancing one narrative of the United States as a threat and advancing another narrative of being a peer of the United States.[193] To facilitate this rhetorical balancing act, the Russian sources we surveyed often pointed to domestic voices within the United States. The authors would note that *U.S.* critiques found a particular U.S. policy or action to be aggressive in nature. This technique was used to a surprising degree to raise—yet simultaneously deflect—concern about U.S. actions in space. But the degree to which Russian sources in *Red Star* are at least following the dialogue taking place within the United States should be accounted for in any effort to shape how Russia perceives U.S. actions in space. Interestingly, we did not find evidence of this same tactic in the Chinese literature surveyed, though Chinese analysts have made similar use of U.S. domestic criticism elsewhere, including in Track II dialogues.

China seeks to strike a different rhetorical balancing act: broadly portraying itself as equal to the United States in overall national power and international status but nonetheless not a military threat, while the United States is a global menace. Even though China aspires to be a great

---

[190] Johan Eriksson and Roman Privalov, "Russian Space Policy and Identity: Visionary or Reactionary?" *Journal of International Relations and Development*, Vol. 24, 2021, pp. 381–382.

[191] Kremlin, "Plenarnoye Zasedaniye Vostochnogo Ekonomicheskogo Foruma" ["Plenary Session of the Eastern Economic Forum"], September 7, 2017.

[192] Eriksson and Privalov, 2021, p. 401.

[193] After the research for this report was completed, Russia signaled its intention to withdraw from cooperation on the International Space Station after 2024. See Joey Roulette, "Russia Signals Space Station Pullout; NASA Says It's Not Official Yet," Reuters, July 26, 2022.

power, and considers itself as such, the Chinese sources surveyed describe it as not yet being the military equal of the United States. Therefore, China sidesteps the need to explain why its own, similar actions should not be considered threatening. This is reflected in Chinese military writings as well. As Pollpeter et al., 2020, found, "Professor Zhou Derong of the PLA Logistics Academy argued [in 2007] that if China were developing ASAT weapons, they would be 'no more than self-defense measures,' and that the United States is falsely depicting itself as a 'victim' to provide an excuse for the development of its own space forces."[194] In contrast to Russia, China was not included in the International Space Station, so its level of space cooperation is lower, leaving Beijing with more rhetorical room to characterize U.S. activities as threatening. The tension for Beijing is reconciling its national self-image and long-term goal of becoming a "world-class military" with its own narrative of being a weaker space power.

Thus, for China, U.S. activities in space—both policies and actions—are aggressive, but China's activities are not aggressive because China is not a military threat on par with the United States. This characterization enables the Chinese MFA to frequently decry what it characterizes as U.S. hypocrisy on militarization. Amid news reports that China conducted tests of a fractional orbital bombardment system in 2021, an MFA spokesperson commented,

> The U.S. is the first country in the world to conduct the research and development of hypersonic weapons. It is now still developing and even spreading hypersonic missile technology and investing trillions of dollars to upgrade its "nuclear triad" force. We have noted that the U.S. expressed concerns over China's normal spacecraft test and play [sic] up the "China threat" theory. Can the U.S. explain to the international community what it intends to do with its development of hypersonic weapons? . . . We urge the U.S. to fully respect other countries' right to develop normal defense capacity, and stop seeking military buildup by hyping up the so-called "military threat" of others.[195]

However, Beijing does not necessarily always hype space-related issues when other concerns about U.S. intentions and capabilities are more pressing. The stated overarching Chinese concern behind China's opposition to the U.S. deployment of THAAD to South Korea was the potential U.S. degradation of China's nuclear capability, the idea being that the radar in South Korea could identify Chinese multiple independent reentry vehicle (MIRV) decoys and distinguish them from

---

[194] Zhang Qiang, "Why Is America Obsessing over China's Missile Tests?" *S&T Daily* [科技日报], December 7, 2007; cited in Pollpeter et al., 2020, p. 59. For another, similar argument justifying China's ASAT program as self-defense and a "legitimate need," according to international law by a non-PLA academic, see Yang Caixia [杨彩霞] and Ai Dun [艾顿], "On the Legality of the Development of ASATs for China" ["论中国发展反卫星武器的合法性"], *Journal of Beijing University of Aeronautics and Astronautics (Social Sciences Edition)* [北京航空航天大学学报（社会科学版）], Vol. 23, No. 2, March 2010.

[195] MFA of the People's Republic of China, "Foreign Ministry Spokesperson Wang Wenbin's Regular Press Conference on October 22, 2021," October 22, 2021b.

actual warheads.[196] This is in contrast to another potential capability that was not raised as a concern in our review of Chinese responses to THAAD: that the THAAD interceptor might also function as a serviceable direct-ascent ASAT weapon.[197]

Modest Support for the Action-Reaction Model

Washington, Beijing, and Moscow all have a general tendency to point to the other side as a threat whose actions require a response, perpetuating a fairly predictable action-reaction cycle, though it can take a long time to unfold in public. Folding in more-recent events, such as the 2018 U.S. National Space Strategy, the creation of the USSF, and the more aggressive rhetoric used by USAF and USSF leaders since 2018, one might draw the conclusion that the United States is simply reacting to the actions of China and Russia. For example, in 2018, U.S. General David L. Goldfein offered the perspective that more-aggressive actions were needed by the United States to ensure that "we are always the predator and never the prey when it comes to competing, deterring and winning in this warfighting domain [space]."[198] Moreover, the President's Budget for fiscal year 2019 offered the largest budget for space since 2003, providing the funding to ensure that the United States remains the "predator" and never becomes the "prey."

Although it is difficult to reach any kind of unequivocal conclusion as to whether Chinese or Russian actions in space have been taken in response to past or more-recent U.S. actions, there are some indications that past U.S. actions have influenced Beijing's and Moscow's perceptions of current U.S. actions, just as past Chinese and Russian actions have influenced Washington. For example, Chinese and Russian actors specifically cite U.S. space activities to support their domestic bureaucratic political arguments for resources, funding, and prioritization of their own space-related activities. Moreover, the Chinese and Russian sources we surveyed consistently characterize the United States as the "first mover," whereas they are simply trying to play catch-

---

[196] For the official Chinese MFA statement, see MFA of the People's Republic of China, "Wang Yi Talks About US's Plan to Deploy THAAD Missile Defense System in ROK," February 13, 2016. For a broader, straightforward explanation of Chinese concerns, see Li Bin, "The Security Dilemma and THAAD Deployment in the ROK," China-US Focus, March 6, 2017. At least some Chinese scholars did publicly affiliate THAAD with SDI, albeit obliquely; see Teng Jianquan, "Why Is China Unhappy with the Deployment of THAAD in the ROK?" China Institute of International Studies, April 1, 2015.

[197] Laura Grego, *The Anti-Satellite Capability of the Phased Adaptive Approach Missile Defense System*, Washington, D.C.: Federation of American Scientists, Winter 2011. For two reviews of Chinese responses, see Ethan Meick and Nargiza Salidjanova, *China's Response to U.S.–South Korean Missile Defense System Deployment and Its Implications*, Washington, D.C.: U.S.-China Economic and Security Review Commission, July 26, 2017; and Michael D. Swaine, "Chinese Views on South Korea's Deployment of THAAD," *China Leadership Monitor*, No. 52, February 2017.

[198] David L. Goldfein, Chief of Staff, U.S. Air Force, remarks, 34th National Space Symposium, Colorado Springs, Colo.: April 17, 2018.

up to counter U.S. efforts to achieve military primacy. This perspective is offered in both external (English) and internal (native-language, primary source) writings.

Yet China and Russia are both highly sensitive to the perception that arms racing is a negative behavior, and both thus emphasize the need to avoid such behavior, *even as they engage in it*, because it makes the world less safe and carries immense costs. Russian authors in *Red Star* and President Putin point to the history of the Soviet Union's arms race with the United States as the cause of the USSR's demise. Russia is *very* conscious of its history in this regard and appears to be keen to not repeat the same mistake. Chinese authors and the Chinese Communist Party also point to the collapse of the USSR, drawing a connection between its collapse and its arms race with the United States, and offer similar calls to avoid arms races. This is an example, though, in which Chinese rhetoric clashes with the reality that Beijing is currently undertaking a massive arms buildup.

To avoid the arms race trap, authors from China and Russia emphasize the utility of asymmetric responses. Sources from both countries appear to view asymmetric responses as being the superior option to arms racing in kind: i.e., seeking out low-cost opportunities to undermine U.S. technical advantages rather than being goaded into trying to match the United States in terms of defense spending and duplicating U.S. programs. In this sense, both China and Russia appear to be acting in response to U.S. actions but seeking to avoid responding in kind and attempting to match U.S. capabilities in the space domain.

The creation of the United States' and China's respective space forces, the USSF and PLASSF, reflects this interwoven action-reaction cycle. Perhaps it should not have come as a surprise that China moved in 2015 to create PLASSF, which is responsible for Chinese military space operations.[199] After its above-noted criticism of the United States for realizing the weaponization of space for the first time, China's 2015 defense white paper further states, "China will keep abreast of the dynamics of outer space, deal with security threats and challenges in that domain, and secure its space assets to serve its national economic and social development, and maintain outer space security."[200] In turn, some in the United States use China's establishment of PLASSF to justify the creation of the USSF.[201] Others, including Lt Gen B. Chance Saltzman, USSF Deputy Chief of Space Operations for Operations, Cyber, and Nuclear, link the USSF's creation even earlier, to the 2007 direct-ascent ASAT test; as Saltzman said in 2020,

---

[199] For more on the creation of PLASSF, see Kevin L. Pollpeter, Michael S. Chase, and Eric Heginbotham, *The Creation of the PLA Strategic Support Force and Its Implications for Chinese Military Space Operations*, Santa Monica, Calif.: RAND Corporation, RR-2058-AF, 2017; and John Costello and Joe McReynolds, *China's Strategic Support Force: A Force for a New Era*, Washington, D.C.: Institute for National Strategic Studies, National Defense University, China Strategic Perspectives 13, 2018.

[200] PRC Information Office of the State Council, 2015.

[201] Dean Cheng, "Does the United States Need a Space Force?" transcript of interview with Michelle Cordero, "Heritage Explains" podcast, Heritage Foundation, July 27, 2018.

> I was on the ops floor in 2007 when the Chinese shot their own satellite down. . . . That was a clarifying event, and I can almost chart from there the establishment of the Space Force because suddenly space was contested. . . . We knew there was other kinds of [space] contesting going on, but that kinetic attack on a satellite really shook the foundations that this is no longer a benign environment, and we started asking the questions about are we properly structured and organized and doing the right kinds of things to be able to maintain our advantage.[202]

A similar phenomenon could be playing out with China's development of its own space plane, which appears to be at least partly in response to the U.S. X-37B. The X-37B is universally assessed by the Chinese sources we surveyed to be a space weapon, or at least a testing platform for future space capabilities.[203] The 2013 AMS SMS states,

> The U.S. is now conducting R&D [research and development] and testing of a group of weapons systems used for conducting space attack and defense operations; its R&D on the X-37B aerospace plane and other space weapons has already realized breakthrough-quality advances, and its capability for executing missile interception in outer space is gradually maturing.[204]

In September 2020, China successfully launched and recovered a space plane, with little explanation except to call it a "reusable experimental spacecraft."[205] Although the specific capabilities and broader intent of China's space plane are still largely uncertain, China has been known to develop military capabilities that have already been acquired by other major powers, largely, if not solely, to better understand the technological aspects of these capabilities but not to deploy them.[206]

Questions for Future Consideration

Taking a step back, it is striking that, in selecting a representative set of pacing events—actions taken by the United States that might plausibly have influenced Chinese and Russian perspectives of the space domain—we broadly "got it right" with respect to Russia. However, some events that mattered to Russia were essentially nonevents in Beijing, and other U.S. space

---

[202] Frank Wolfe, "Space Force Official Suggests U.S. Developing Offensive Capabilities in Space," *Defense Daily*, October 19, 2020.

[203] Dean Cheng, "When the Chinese Look at the US X-37B, They See the Future of Space-Based Attack," *Foreign Policy*, October 23, 2014.

[204] AMS Military Strategy Department, 2013, p. 183.

[205] "China Launches Reusable Experimental Spacecraft," Xinhua, September 7, 2020. For U.S. government discussion of the space plane, see Office of the Secretary of Defense, 2021, p. 66.

[206] Jonathan Ray, *Red China's "Capitalist Bomb": Inside the Chinese Neutron Bomb Program*, Washington, D.C.: Institute for National Strategic Studies, National Defense University, China Strategic Perspectives 8, January 1, 2015; and Wu Riqiang, "How China Practices and Thinks About Nuclear Transparency," in Li Bin and Tong Zhao, eds., *Understanding Chinese Nuclear Thinking*, Washington, D.C.: Carnegie Endowment for International Peace, 2016, p. 237.

activities received a great deal of attention in China, such as the Schriever [施里弗] space wargame series.[207] Interestingly, these U.S. space wargames were *not* mentioned in the Russian sources that we analyzed. This result suggests that the diplomatic history and cultural understanding between the United States and the USSR/Russia was perhaps more formative than might be fully appreciated. This history appears to allow the United States to manage its relationship more effectively with Russia—or at least to understand how its policies and actions might be perceived by Russia.[208] Such a history and cultural understanding is nearly absent in the U.S. relationship with China. Our analysis suggests that U.S. policymakers should be relatively modest regarding the U.S. ability to anticipate or manage Chinese perceptions of U.S. actions in the space domain.

Lastly, we raise two of many potential questions for future consideration that are based on the research in this report:

- As China and Russia increase cooperation in space, such as remote sensing, satellites, a potential joint lunar base, and even missile early warning, will this drive a further convergence of a common perspective of U.S. space activities and greater coordination on the international stage?[209] Beijing and Moscow have already been cooperating for decades in the Conference on Disarmament to oppose U.S. ballistic missile defense and to try to get the Treaty on the Prevention of the Placement of Weapons in Outer Space and of the Threat or Use of Force Against Outer Space Objects passed, which would limit space weaponization. How might this cooperation evolve in the future?
- Alternatively, just as the modernization of China's military continues to mean that it is catching up to the U.S. military, this also means that it is matching or even surpassing Russian military capabilities in some areas. Recognizing the potential for this current period of cooperation to abate because of Russian concerns over a Chinese military threat, it will be important to consider how Russia might view Chinese space capabilities. Our cursory review found that Russia appears to cite U.S. literature to circuitously criticize Chinese military capabilities. This triangular relationship between Washington, Moscow, and Beijing in space dates back decades; a declassified 1986 CIA report on contemporary Chinese reactions to SDI found that Chinese analysts "stress that Moscow's goal in opposing SDI is not to prevent the deployment of space-based weapons, but to delay the U.S. effort until the Soviet Union is in a better position to compete . . . . Of greater concern is the fear that the Soviets will deploy an SDI system that will neutralize China's small nuclear deterrent."[210]

---

[207] For some PLA sources on Schriever and other U.S. space wargames, see Yuan, 2001, pp. 26–29; Chang, 2005; Jiang, 2013, pp. 12, 263; and Peng, Lv, and Lu, 2019, pp. 59–64.

[208] For more on this, see John Lewis Gaddis, "Learning to Live with Transparency: The Emergence of a Reconnaissance Satellite Regime," in John Lewis Gaddis, *The Long Peace: Inquiries into the History of the Cold War*, New York: Oxford University Press, 1987, pp. 195–214; and James Clay Moltz, *The Politics of Space Security: Strategic Restraint and the Pursuit of National Interests*, Stanford, Calif.: Stanford University Press, 2019.

[209] For more, see Richard Weitz, "Sino-Russian Cooperation in Outer Space: Taking Off?" *China Brief*, Vol. 20, No. 21, December 6, 2020.

[210] CIA, 1986.

Abbreviations

ABM	Anti-Ballistic Missile
AEU	Aerospace Engineering University
AMS	Academy of Military Science
ASAT	anti-satellite
BMDO	Ballistic Missile Defense Office
CIA	Central Intelligence Agency
CNKI	China National Knowledge Infrastructure
DoD	U.S. Department of Defense
GPS	Global Positioning System
GSSAP	Geosynchronous Space Situational Awareness Program
ICBM	intercontinental ballistic missile
MFA	Ministry of Foreign Affairs
MIRACL	Mid-Infrared Advanced Chemical Laser
NDU	National Defense University
PLA	People's Liberation Army
PLASSF	People's Liberation Army Strategic Support Force
PRC	People's Republic of China
SDI	Strategic Defense Initiative
SM-3	Standard Missile-3
SMS	*Science of Military Strategy*
SSD	Space Systems Department
THAAD	Terminal High Altitude Area Defense
TMDP	Theater Missile Defense Program
USAF	U.S. Air Force
USSF	U.S. Space Force
USSR	Union of Soviet Socialist Republics

Bibliography

Unless otherwise indicated, the authors of this report provided the translations of bibliographic details for the non-English sources included in this report. To support conventions for alphabetizing, sources in Chinese are introduced with and organized according to their English translations. The original rendering in Chinese appears in brackets after the English translation. Bibliographical details in Russian are introduced with and organized according to their transliteration into the Latin alphabet.

Academy of Military Science [军事科学院], ed., *Science of Military Strategy* [战略学], 1st ed., Beijing: Academy of Military Science Press [军事科学出版社], 1987.

Academy of Military Science Military Strategy Department [军事科学院战略研究部], ed., *Science of Military Strategy* [战略学], 2nd ed., Beijing: Academy of Military Science Press [军事科学出版社], 2001.

Academy of Military Science Military Strategy Department [军事科学院战略研究部], ed., *Science of Military Strategy* [战略学], 3rd ed., Beijing: Academy of Military Science Press [军事科学出版社], 2013.

Air Force Doctrine Document 2-2.1, *Counterspace Operations*, Washington, D.C.: Headquarters Air Force Doctrine Center, August 2, 2004.

Allen, Kenneth, and Mingzhi Chen, *The People's Liberation Army's 37 Academic Institutions*, Washington, D.C.: China Aerospace Studies Institute, 2020.

AMS—*See* Academy of Military Science.

Andreev, Dmitriy, "Gospodstvuyushchiye Vysoty Kosmicheskikh Voysk" ["Superior Heights of the Space Forces"], *Krasnaia Zvezda* [*Red Star*], No. 14, 2008.

Antsupov, O. I., and A. S. Zhikharev, "Analiz Osnovnykh Kontseptual'nykh Podkhodov k Sozdaniyu Sistem Strategicheskoy PRO SSHA i Rossiyskoy Federatsii" ["Analysis of the Main Conceptual Approaches to the Creation of Strategic Missile Defense Systems of the USA and the Russian Federation"], *Voennaia Mysl'* [*Military Thought*], No. 6, 2015.

Boese, Wade, "U.S. Withdraws from ABM Treaty; Global Response Muted," *Arms Control Today*, July/August 2002.

Bowe, Alexander, *China's Pursuit of Space Power Status and Implications for the United States*, Washington, D.C.: U.S.-China Economic and Security Review Commission, April 11, 2019.

Bradsher, Keith, "China Criticizes U.S. Missile Strike," *New York Times*, February 22, 2008.

Buzan, Barry, and Eric Herring, *The Arms Dynamic in World Politics*, Boulder, Colo.: Lynne Rienner Publishers, 1998.

Central Intelligence Agency, "Views of Chinese Military and Civilian Analysts on the Strategic Defense Initiative," January 1986, declassified May 27, 2011.

Chang Xianqi [常显奇], *Military Astronautics* [军事航天学], 2nd ed., Beijing: National Defense Industries Press [国防工业出版社], 2005.

Chel'tsov, B. F., "Voprosy Vozdushno-Kosmicheskoy Oborony v Voyennoy Doktrine" ["Aerospace Defense Issues in the Russian Military Doctrine"], *Voennaia Mysl'* [*Military Thought*], No. 4, April 2007.

Chen Jie [陈杰], Pan Feng [潘峰], and Su Tongling [苏同领], "EHF Satellite Communication System of the U.S. Army" ["美国天基太空监视系统"], *National Defense Technology* [国防科技], 2011, pp. 67–70.

Cheng, Dean, "When the Chinese Look at the US X-37B, They See the Future of Space-Based Attack," *Foreign Policy*, October 23, 2014.

———, "Does the United States Need a Space Force?" transcript of interview with Michelle Cordero, "Heritage Explains" podcast, Heritage Foundation, July 27, 2018.

Cheng, Dean, Peter Garretson, Namrata Goswami, James Lewis, Bruce W. MacDonald, Kazuto Suzuki, Brian C. Weeden, and Nicholas Wright, *Outer Space; Earthly Escalation? Chinese Perspectives on Space Operations and Escalation*, ed. Nicholas Wright, Washington, D.C.: Joint Staff, U.S. Department of Defense, August 2018.

Cheng Qun [程群], "On the Adjustment of the Obama Administration's Space Policy" ["浅析奥巴马政府太空政策的调整"], *Contemporary International Relations* [现代国际关系], June 2010.

"China Launches Reusable Experimental Spacecraft," Xinhua, September 7, 2020.

CIA—*See* Central Intelligence Agency.

Clark, Stephen, "Air Force General Reveals New Space Surveillance Program," Space.com, March 3, 2014.

Commission to Assess United States National Security Space Management and Organization, *Report of the Commission to Assess United States National Security Space Management and Organization*, Washington, D.C., January 11, 2001.

Costello, John, and Joe McReynolds, *China's Strategic Support Force: A Force for a New Era*, Washington, D.C.: Institute for National Strategic Studies, National Defense University, China Strategic Perspectives 13, 2018.

Defense Intelligence Agency, *Challenges to Security in Space*, Arlington, Va., 2019.

DoD—*See* U.S. Department of Defense.

Dokuchayev, Anatoliy, "Zametki Voyennogo Obozrevatelya. 'Brillianty' Nad Rossiyey i Pogreba Pod Ney" ["Notes of a Military Observer. 'Diamonds' over Russia and a Cellar Under It"], *Krasnaia Zvezda* [*Red Star*], No. 47, 1992.

———, "Rakety i Kosmos. Okhota Sredi Zvezd" ["Rockets and Space. Hunting Among the Stars"], *Krasnaia Zvezda* [*Red Star*], No. 264, 1999.

———, "Tochku Vstrechi Izmenit' Nel'zya. Zavtra Protivoraketchiki Rossii Otmetyat Yubiley" ["The Meeting Point Cannot Be Changed. Tomorrow, Russia's Missile Defense Forces Will Celebrate Their Anniversary"], *Krasnaia Zvezda* [*Red Star*], No. 42, 2001.

Dontsov, Viktor, "'Tri Plyus Tri.' Pod Takim Nezateylivym Nazvaniyem Skryvayetsya Amerikanskaya Programma Sozdaniya Natsional'noy PRO" ["'Three Plus Three.' The American Program for the Creation of a National Missile Defense System Is Hidden Under This Unpretentious Name"], *Krasnaia Zvezda* [*Red Star*], No. 260, 1998.

Drea, Edward J., Ronald H. Cole, Walter S. Poole, James F. Schnabel, Robert J. Watson, and Willard J. Webb, *History of the Unified Command Plan, 1946–2012*, Washington, D.C.: Joint History Office, Office of the Chairman of the Joint Chiefs of Staff, 2013.

Eriksson, Johan, and Roman Privalov, "Russian Space Policy and Identity: Visionary or Reactionary?" *Journal of International Relations and Development*, Vol. 24, 2021, pp. 381–407.

Falaleyev, Mikhail, "Ministr Oborony Rossiyskoy Federatsii Sergey Ivanov: Glavnyy Kriteriy - Bezopasnost' Rossii" ["Defense Minister of the Russian Federation Sergei Ivamov: The Main Criterion Is the Security of Russia"], *Krasnaia Zvezda* [*Red Star*], No. 136, 2001.

Falichev, Oleg, "Vizit. Ministr Oborony Rossii v Belgrade" ["Visit. Russian Defense Minister in Belgrade"], *Krasnaia Zvezda* [*Red Star*], No. 270, 1999.

Fan Gaoyue [樊高月] and Gong Xuping [宫旭平], "The Development and Evolution of U.S. Space Strategic Thought (Part 1)" ["美国太空战略思想的发展与演变(上)"], *National Defense* [国防], February 2016a.

———, "The Development and Evolution of U.S. Space Strategic Thought (Part 2)" ["美国太空战略思想的发展与演变(下)"], *National Defense* [国防], March 2016b.

Feng Songjiang [丰松江] and Dong Zhenghong [董正宏], *Space, the Future Battlefield: New Situation and New Trends in the U.S. Militarization of Space* [太空未来战场: 美国太空军事化新态势新走向], Beijing: Current Affairs Press [时事出版社], 2021.

Gaddis, John Lewis, "Learning to Live with Transparency: The Emergence of a Reconnaissance Satellite Regime," in John Lewis Gaddis, *The Long Peace: Inquiries into the History of the Cold War*, New York: Oxford University Press, 1987, pp. 195–214.

Gaoyang Yuxi [高杨予兮], "The Historical Evolution of U.S. Space Deterrence Strategy" ["美国太空威慑战略的历史演进"], *International Study Reference* [国际研究参考], June 2017, pp. 27–34.

———, "Adjustment of US Space Deterrence Strategy and Its Impact" ["美国太空威慑战略调整及其影响"], *Peace and Development* [和平与发展], March 2018, pp. 116–135.

Gaoyang Yuxi [高杨予兮] and Ke Long [柯隆], "New Trends of U.S. Space Cooperation Policy" ["美国太空合作政策新动向"], *International Study Reference* [国际研究参考], June 2014, pp. 1–45.

Garver, John W., "China's Response to the Strategic Defense Initiative," *Asian Survey*, Vol. 26, No. 11, November 1986, pp. 1220–1239.

Glaser, Bonnie S., and Banning N. Garrett, "Chinese Perspectives on the Strategic Defense Initiative," *Problems with Communism*, Vol. 35, No. 2, March–April 1986, pp. 28–44.

Goldfein, David L., Chief of Staff, U.S. Air Force, remarks, 34th National Space Symposium, Colorado Springs, Colo., April 17, 2018. As of November 11, 2021: https://www.af.mil/Portals/1/documents/csaf/CSAF%20Remarks-NationalSpaceSymposium.pdf?ver=2018-04-20-130722-347

Gol'ts, Aleksandr, "PRO Natselena... v Klintona" ["Missile Defense Targets... Clinton"], *Krasnaia Zvezda* [*Red Star*], No. 146, 1996.

Graham, Daniel O., *Confessions of a Cold Warrior*, Fairfax, Va.: Preview Press, 1995.

Grego, Laura, *The Anti-Satellite Capability of the Phased Adaptive Approach Missile Defense System*, Washington, D.C.: Federation of American Scientists, Winter 2011.

Guo Jun [郭俊], "Construction of U.S. Space Deterrence Force Revealed in 'Schriever' Exercises" ["从'施里弗'演习看美军太空威慑力量构建"], *National Defense Technology* [国防科技], Vol. 36, No. 1, February 2015, pp. 68–89.

He Qingsong [何奇松], "Fragile High Frontier: The Strategic Dilemma of U.S. Space Deterrence in the Post–Cold War Era" ["脆弱的高边疆:后冷战时代 美国太空威慑的战略困境"], *Social Sciences in China* [中国社会科学], April 2012, pp. 183–204.

———, "Interactions Between China and the United States in the Area of Space Security" ["中美两国太空安全领域的互动"], *Journal of International Security Studies* [国际安全研究], September 2017, pp. 26–52.

Hill, Jon A., and Michelle C. Atkinson, "Department of Defense Press Briefing on the President's Fiscal Year 2022 Defense Budget for the Missile Defense Agency," press conference transcript, U.S. Department of Defense, May 28, 2021.

Hu Guangzheng [胡光正] and Kan Nan [阚南], "To Seize the Power of Heaven, Who Is the Final Winner?" ["夺取制天权, 谁是最后的赢家?"], *Aerospace Knowledge* [航空知识], December 2006.

Hu Xujie [胡绪杰] and Zhang Zhifeng [张志峰], "Research on the Development of U.S. Aerospace Forces" ["美国航天力量发展研究"], *O. I. Automation* [兵工自动化], Vol. 26, No. 9, 2007.

Hui Zhang, "The U.S. Weaponization of Space: Chinese Perspectives," presentation at Nuclear Policy Research Institute Conference, *Full Spectrum Dominance: The Impending Weaponization of Space*, Warrenton, Va., May 16–17, 2005.

———, "Chinese Perspectives on Space Weapons," in Pavel Podvig and Hui Zhang, eds., *Russian and Chinese Responses to U.S. Military Plans in Space*, Cambridge, Mass.: American Academy of Arts and Sciences, January 2008, pp. 31–77.

Huo Mu [火木], "What Is the True Intention of the United States to Destroy Runaway Satellites with Sea-Based Missiles?" ["美国用海基导弹摧毁失控卫星真实意图何在?"], *Modern Navy* [当代海军], April 2008.

"In Test, Military Hits Satellite Using a Laser," *New York Times*, October 21, 1997.

"Inostrannaya Voyennaya Khronika" ["Foreign Military Chronicle"], *Krasnaia Zvezda* [*Red Star*], No. 116, 2012a.

"Inostrannaya Voyennaya Khronika" ["Foreign Military Chronicle"], *Krasnaia Zvezda* [*Red Star*], No. 215, 2012b.

"Inostrannaya Voyennaya Khronika" ["Foreign Military Chronicle"], *Krasnaia Zvezda* [*Red Star*], No. 86, 2015.

Jiang Lianju [姜连举], ed., *Lectures on the Science of Space Operations* [空间作战学教程], Beijing: Military Science Press [军事科学出版社], 2013.

Johnson, Nicholas L., "Operation Burnt Frost: A View from Inside," *Space Policy*, Vol. 56, May 2021.

Johnston, Alastair Iain, *China and Arms Control: Emerging Issues and Interests in the 1980s*, Ottawa, Canada: Canadian Centre for Arms Control and Disarmament, 1986.

Kennedy, John F., "Address Before the General Assembly of the United Nations," September 25, 1961. As of January 27, 2022: https://www.jfklibrary.org/archives/other-resources/john-f-kennedy-speeches/united-nations-19610925

Khan, Joseph, "China Confirms Test of Anti-Satellite Weapon," *New York Times*, January 23, 2007.

Khromov, Gennadiy, "Tochka Zreniya. Uvazhayut Tekh, Kto Dumayet O Blage Strany" ["Point of View. Those Who Care About the Good of Their Countries Are Respected"], *Krasnaia Zvezda* [*Red Star*], Nos. 243–244, 1992.

Khryapin, Aleksandr, and Oleg Pyshnyy, "V 'Avangarde' Progressa" ["In the 'Vanguard' of Progress"], *Krasnaia Zvezda* [*Red Star*], No. 11, 2019.

Kirsanov, Dmitriy, "Kosmicheskiye Orbity Pentagona" ["Pentagon Space Orbits"], *Krasnaia Zvezda* [*Red Star*], No. 66, 2012.

Kislyakov, Andrey, "Pentagon Gotovitsya k Vedeniyu Boyevykh Deystviy v Kosmose" ["The Pentagon Prepares to Conduct Hostilities in Space"], *Krasnaia Zvezda* [*Red Star*], No. 24, 2019a.

———, "SSHA Zanosyat Nad Planetoy Damoklov Mech" ["The United States Raises the Sword of Damocles over the Planet"], *Krasnaia Zvezda* [*Red Star*], No. 46, 2019b.

———, "V Kosmose Stanovitsya Nebezopasno" ["The Space Becomes Unsafe"], *Krasnaia Zvezda* [*Red Star*], No. 45, 2020.

Konopatov, S. N., and Ye. A. Starozhuk, "Kosmicheskiye Sistemy v Novoy Srede Bezopasnosti" ["Space Systems in a New Security Environment"], *Voennaia Mysl'* [*Military Thought*], No. 1, January 2019.

"Kosmicheskiy Machizm Opasen Dlya Chelovechestva" ["Space Machismo Is Dangerous for the Humanity"], *Krasnaia Zvezda* [*Red Star*], No. 23, 2007.

Kozin, Vladimir, "Aziatskiy 'Dovesok' k YEVROPRO" ["Asian 'Add-on' to European Missile Defense"], *Krasnaia Zvezda* [*Red Star*], No. 69, 2008.

———, "Pol'sha, Rumyniya... Kto Sleduyushchiy?" ["Poland, Romania... Who's Next?"], *Krasnaia Zvezda* [*Red Star*], No. 21, 2010.

———, "V Dukhe 'Zvozdnykh Voyn'" ["Inspired by the 'Star Wars'"], *Krasnaia Zvezda* [*Red Star*], No. 3, 2016.

———, "SSHA Rassmatrivayut Kosmos Kak Budushcheye 'Pole Boya'" ["US Sees Space as a Future 'Battlefield'"], *Krasnaia Zvezda* [*Red Star*], No. 29, 2018.

Kremlin, "Interv'yu teleradiokompanii 'Mir,'" ["Interview with the Television and Radio Company 'Mir'"], April 12, 2017. As of October 1, 2021: http://kremlin.ru/events/president/news/54271

———, "Plenarnoye Zasedaniye Vostochnogo Ekonomicheskogo Foruma" ["Plenary Session of the Eastern Economic Forum"], September 7, 2017. As of October 22, 2021: http://kremlin.ru/events/president/news/55552

———, "Vstrecha s pomoshchnikom Prezidenta SSHA po natsbezopasnosti Dzhonom Boltonom" ["Meeting with Assistant to the President of the United States for National Security Affairs John Bolton"], October 23, 2018. As of October 1, 2021: http://kremlin.ru/events/president/news/58880

Kulacki, Gregory, *Anti-Satellite (ASAT) Technology in Chinese Open-Source Publications*, Cambridge, Mass.: Union of Concerned Scientists, July 1, 2009.

Kulacki, Gregory, and Jeffrey G. Lewis, "Understanding China's Antisatellite Test," *Nonproliferation Review*, Vol. 15, No. 2, 2008, pp. 335–347.

Kuzar', Vladimir, "Voyenno-Politicheskoye Obozreniye. Cherez Preniya k Zvezdam" ["Military-Political Review. Through Debate to the Stars"], *Krasnaia Zvezda* [*Red Star*], No. 111, 2004.

———, "Voyenno-Politicheskoye Obozreniye. Pole Boya – Vselennaya" ["Military-Political Review. The Universe Is the Battlefield"], *Krasnaia Zvezda* [*Red Star*], No. 60, 2006.

———, "Kosmos Nadevayet Kamuflyazh" ["The Space Puts on the Camouflage"], *Krasnaia Zvezda* [*Red Star*], No. 37, 2008a.

———, "SSHA: Problemy Rastut, Ambitsii Te Zhe" ["The USA: Growing Problems, Same Ambitions"], *Krasnaia Zvezda* [*Red Star*], No. 15, 2008b.

Lewis, John Wilson, and Hua Di, "China's Ballistic Missile Programs: Technologies, Strategies, Goals," *International Security*, Vol. 17, No. 2, Fall 1992, pp. 5–40.

Li Bin, "The Security Dilemma and THAAD Deployment in the ROK," China-US Focus, March 6, 2017.

Li Daguang [李大光], "Look at the Development of Anti-Satellite Weapons from the United States' 'Missile-to-Satellite'" ["由美国'导弹打卫星'看其反卫星武器发展"], *Technology Foundation of National Defence* [国防技术基础], July 2008, pp. 40–45.

———, "Looking at the Development of Anti-Satellite Weapons from the U.S. 'Missile-to-Satellite'" ["由美国'导弹打卫星'看其反卫星武器发展"], *Defense Industry Conversion in China* [中国军转民], July 2010, pp. 62–66.

Li Shuang [李爽], *A Study on the Code of Conduct in Outer Space from the Perspective of Governance Theory* [治理理论视阈下的外空活动行为准则研究], master's thesis, Changsha, China: PLA National University of Defense Technology [国防科学技术大学], 2014.

Li Yan [李焱], "The Latest Development Trend and Analysis of U.S. Space Weapons" ["美国太空武器最新发展动向及分析"], *Space International* [国际太空], 2008, pp. 18–24.

Long Xuedan [龙雪丹] and Qu Jing [曲晶], "Review of World Launch Vehicles in 2012" ["2012年世界航天运载器发展回顾"], *Missiles and Space Vehicles* [导弹与航天运载技术], February 2013, pp. 31–37.

Luo Xi [罗曦], "The Adjustments of U.S. Strategic Deterrence System and Their Implications to Sino-US Strategic Stability" ["美国战略威慑体系的调整与中美战略稳定性"], *Journal of International Relations* [国际关系研究], June 2017.

Lyashchenko, Aleksey, "Voyenno-Politicheskoye Obozreniye. Ambitsioznyye Plany Washingtona" ["Military-Political Review. Washington's Ambitious Plans"], *Krasnaia Zvezda* [*Red Star*], No. 105, 2002.

———, "Voyenno-Politicheskoye Obozreniye. Reanimatsiya Gonki Yadernykh Vooruzheniy" ["Military-Political Review. Reanimation of the Nuclear Arms Race"], *Krasnaia Zvezda* [*Red Star*], No. 147, 2005.

Markushin, Vadim, "Marsiada Busha" ["Bush's Martian Project"], *Krasnaia Zvezda* [*Red Star*], No. 4, 2004.

Martymyanov, Aleksey, and Aleksandr Dolinin, "Nevidannyy Proryv" ["The Unseen Breakthrough"], *Krasnaia Zvezda* [*Red Star*], No. 239, 2006.

Medeiros, Evan, *Reluctant Restraint: The Evolution of China's Nonproliferation Policies and Practices, 1980–2004*, Stanford, Calif.: Stanford University Press, 2007.

Medvedev, Sergey, "Kosmicheskiy Proryv Obamy" ["Obama's Space Breakthrough"], *Krasnaia Zvezda* [*Red Star*], No. 66, 2010.

Meick, Ethan, and Nargiza Salidjanova, *China's Response to U.S.–South Korean Missile Defense System Deployment and Its Implications*, Washington, D.C.: U.S.-China Economic and Security Review Commission, July 26, 2017.

MFA—*See* Ministry of Foreign Affairs.

Mikhaylov, Nikolay, "K Voprosu o Dogovore po PRO 1972 g" ["On the 1972 ABM Treaty"], Krasnaia Zvezda [*Red Star*], No 273, December 30, 1999.

Mil, Seumas, "Bez Kommentariyev. Mozhno Li Pobedit' Soyedinennyye Shtaty?" ["No Comments. Can the United States Be Defeated?"], *Krasnaia Zvezda* [*Red Star*], No. 35, February 22, 2002.

Ministry of Foreign Affairs of the People's Republic of China, "Wang Yi Talks About US's Plan to Deploy THAAD Missile Defense System in ROK," February 13, 2016.

———, "Statements by Chinese Delegation at the 55th Session of the Scientific and Technical Subcommittee of the Committee on the Peaceful Uses of Outer Space," February 14, 2018.

———, "Foreign Ministry Spokesperson Wang Wenbin's Regular Press Conference on October 19, 2021," October 19, 2021a.

———, "Foreign Ministry Spokesperson Wang Wenbin's Regular Press Conference on October 22, 2021," October 22, 2021b.

Missile Defense Agency, U.S. Department of Defense, "Aegis Ballistic Missile Defense: One-Time Mission: Operation Burnt Frost," undated. As of October 20, 2021: https://web.archive.org/web/20120214031001/http://www.mda.mil/system/aegis_one_time_mission.html

Moltz, James Clay, *The Politics of Space Security: Strategic Restraint and the Pursuit of National Interests*, Stanford, Calif.: Stanford University Press, 2019.

National Air and Space Intelligence Center, *Competing in Space*, Dayton, Ohio: Wright-Patterson Air Force Base, December 2018.

National Science and Technology Council, "Fact Sheet: National Space Policy," The White House, September 19, 1996. As of October 20, 2021: https://prd-wret.s3.us-west-2.amazonaws.com/assets/palladium/production/atoms/files/1996%20National%20Space%20Policy.pdf

Oelrich, Ivan, "Transcript: Spy Satellite Shootdown," *Washington Post*, February 20, 2008. As of October 20, 2021: https://www.washingtonpost.com/wp-dyn/content/discussion/2008/02/20/DI2008022001328.html

Office of the Secretary of Defense, *Military and Security Developments Involving the People's Republic of China*, Washington, D.C.: U.S. Department of Defense, 2021.

Pechurov, Sergey, "Formuly Akademika Kokoshina, Ili Kak Obespechit' Intellektual'noye Prevoskhodstvo Rossii" ["Formulas by Academician Kokoshin, or How to Ensure Russia's Intellectual Superiority"], *Krasnaia Zvezda* [*Red Star*], No. 69, 2014.

Peng Guangqian [彭光谦] and Yao Youzhi [姚有志], *Science of Military Strategy*, Beijing: Military Science Publishing House [军事科学出版社], 2005 (translation of 2001 Chinese version).

Peng Huiqiong [彭辉琼], Lv Jiuming [吕久明], and Lu Jiangong [路建功], "Analysis of the Main Results of American Space Combat Exercises" ["美国太空作战演习主要成果探析"], *Aerospace Electronic Countermeasures* [航天电子对抗], Vol. 35, No. 2, 2019, pp. 59–64.

People's Republic of China Information Office of the State Council, "China: Arms Control and Disarmament," November 1995.

———, "China's National Defense in 2002," Xinhua, December 9, 2002.

———, "China's Military Strategy," Xinhua, May 2015.

———, "China's National Defense in the New Era," Xinhua, July 24, 2019.

Permanent Mission of the People's Republic of China to the United Nations, "Document of the People's Republic of China Pursuant to UNGA Resolution 75/36 (2020)," April 30, 2021.

Pogorelyy, M., "Lidirovat' Ili Dogonyat'?" ["Lead or Catch Up?"], *Krasnaia Zvezda* [*Red Star*], No. 34, 1992.

Pogorelyy, Mikhail, "Ot Zvezd, Cherez Politicheskiye Ternii, - Na Zemlyu" ["From the Stars, Through Political Hardships, to the Earth"], *Krasnaia Zvezda* [*Red Star*], Nos. 108–109, 1993.

Pollpeter, Kevin, "The Chinese Vision of Space Military Operations," in James Mulvenon and David Finkelstein, eds., *China's Revolution in Doctrinal Affairs: Emerging Trends in the Operational Art of the Chinese People's Liberation Army*, Alexandria, Va.: CNA Corporation, 2005, pp. 329–369.

Pollpeter, Kevin, Eric Anderson, Jordan Wilson, and Fan Yang, *China Dream, Space Dream: China's Progress in Space Technologies and Implications for the United States*, Washington, D.C.: U.S.-China Economic and Security Review Commission, March 2, 2015.

Pollpeter, Kevin L., Michael S. Chase, and Eric Heginbotham, *The Creation of the PLA Strategic Support Force and Its Implications for Chinese Military Space Operations*, Santa Monica, Calif.: RAND Corporation, RR-2058-AF, 2017. As of January 21, 2022: https://www.rand.org/pubs/research_reports/RR2058.html

Pollpeter, Kevin, Timothy Ditter, Anthony Miller, and Brian Waidelich, *China's Space Narrative: Examining the Portrayal of the US-China Space Relationship in Chinese Sources and Its Implications for the United States*, Montgomery, Ala.: China Aerospace Studies Institute, 2020.

Polyakova, Anna, "Imperativy 'Tsarya Gory'" ["The Imperatives of the 'King of the Hill'"], *Krasnaia Zvezda* [*Red Star*], No. 167, 2015.

———, "Vyvod Oruzhiya v Kosmos Otkroyet Yashchik Pandory" ["Putting Weapons into Space Will Open Pandora's Box"], *Krasnaia Zvezda* [*Red Star*], No. 15, 2016.

Ponomarev, Manki, "Amerikanskikh Yadershchikov Po-Prezhnemu Zhdet Svetloye Budushcheye" ["American Nuclear Scientists Still Have a Bright Future"], *Krasnaia Zvezda* [*Red Star*], No. 174, August 4, 1992.

———, "SOI – Naslediye 'Kholodnoy Voyny'" ["SDI – A Legacy of the Cold War"], *Krasnaia Zvezda* [*Red Star*], No. 65, 1993a.

———, "Voyennoye Obozreniye. SOI: Zhizn' Posle Smerti?'" ["Military Review. SDI: Life After Death?"], *Krasnaia Zvezda* [*Red Star*], No. 168, 1993b.

PRC—*See* People's Republic of China.

President of the Russian Federation, *Voyennaya Doktrina Rossiyskoy Federatsii* [*Military Doctrine of the Russian Federation*], December 25, 2014. As of October 19, 2021: http://static.kremlin.ru/media/events/files/41d527556bec8deb3530.pdf

Press-Tsentr MID RF [Press Center of the Ministry of Foreign Affairs of the Russian Federation], "Briefing No 72," Integrum, October 21, 1997.

———, "O Pravovom Statuse Dogovora Mezhdu Rossiyey i SSHA o Dal'neyshemsokrashchenii i Ogranichenii Strategicheskikh Nastupatel'nykh Vooruzheniy" ["On the Legal Status of the Treaty Between Russia and the United States on the Further Reduction and Limitation of Strategic Offensive Arms"], June 14, 2002.

"Programma SOI Stalkivayetsya s Ser'yeznymi Problemami" ["SDI Program Faces Major Challenges"], *Krasnaia Zvezda* [*Red Star*], No. 60, 1992.

"Protivoraketnaya Oborona: Ukrepleniye Strategicheskoy Stabil'nosti Ili Novyy Vitok Gonki Vooruzheniy?" ["Missile Defense: Strengthening Strategic Stability or a New Round of the Arms Race?"], *Krasnaia Zvezda* [*Red Star*], No. 118, 2000.

"Protivoraketnyy Shchit Dlya Yevropy" ["Missile Defense Shield for Europe"], *Krasnaia Zvezda* [*Red Star*], No. 222, 2006.

Public Law 106-65, National Defense Authorization Act for Fiscal Year 2000, October 5, 1999.

Qu Jing [曲晶] and Ge Shanshan [葛姗姗], "Development of Foreign Launch Vehicles in 2010" ["2010 年世界航天运载器发展回顾"], *Missiles and Space Vehicles* [导弹与航天运载技术], February 2011, pp. 29–35.

Quam, Erik, *Examining China's Debate on Military Space Programs: Was the ASAT Test Really a Surprise?* Washington, D.C.: Nuclear Threat Initiative, January 31, 2007.

Rathjens, George W., "The Dynamics of the Arms Race," *Scientific American*, Vol. 220, No. 4, April 1969, pp. 15–25.

Ray, Jonathan, *Red China's "Capitalist Bomb": Inside the Chinese Neutron Bomb Program*, Washington, D.C.: Institute for National Strategic Studies, National Defense University, China Strategic Perspectives 8, January 1, 2015.

Rebrov, Mikhail, "Sekrety Sekretnykh Sluzhb. DEZA, Ili Shirokiy Front 'Psikhologicheskoy Voyny'" ["Secrets of the Secret Services. DISINFO, or the Wide Frontier of 'Psychological Warfare'"], *Krasnaia Zvezda* [*Red Star*], 1995.

Renshon, Jonathan, "Assessing Capabilities in International Politics: Biased Overestimation and the Case of the Imaginary 'Missile Gap,'" *Journal of Strategic Studies*, Vol. 32, No. 1, February 2009, pp. 115–147.

Report on Scientific Developments in the Aerospace Field: 2017 [航天领域科技 发展报告], Beijing: National Defense Industry Press [国防工业出版社], April 2018.

Roberts, Brad, *China and Ballistic Missile Defense: 1955 to 2002 and Beyond*, Alexandria, Va.: Institute for Defense Analyses, September 2003.

———, *On Theories of Victory, Red and Blue*, Livermore, Calif.: Center for Global Security Research, Lawrence Livermore National Laboratory, No. 7, June 2020.

Roulette, Joey, "Russia Signals Space Station Pullout; NASA Says It's Not Official Yet," Reuters, July 26, 2022.

Ruchkin, Viktor, "Pentagon Nachinayet 'Zvezdnyye Voyny'" ["Pentagon Starts the 'Star Wars'"], *Krasnaia Zvezda* [*Red Star*], No. 27, 2008.

———, "Bez 'Zvezdnykh Voin'" ["No Star Wars"], *Krasnaia Zvezda* [*Red Star*], No. 175, 2009.

———, "V Kosmose Tainstvennyy Chelnok" ["The Mysterious Shuttle in Space"], *Krasnaia Zvezda* [*Red Star*], No. 72, 2010.

Runov, Vladislav, "Ne Narushat' Strategicheskoy Stabil'nosti" ["Do Not Disrupt Strategic Stability"], *Krasnaia Zvezda* [*Red Star*], No. 86, 2001.

Ryzhkov, Sergey, "Chto Budet s Dogovorom Po Otkrytomu Nebu?" ["What Will Happen to the Treaty on Open Skies?"], *Krasnaia Zvezda* [*Red Star*], No. 77, 2020.

Satarov, V., "Stremitel'nyy Global'nyy Udar" ["Prompt Global Strike"], *Krasnaia Zvezda* [*Red Star*], No. 168, 2007.

Security Council of the Russian Federation, *Voyennaya Doktrina Rossiyskoy Federatsii* [*The Military Doctrine of the Russian Federation*], No. 706, Moscow, April 21, 2000.

———, *Voyennaya Doktrina Rossiyskoy Federatsii* [*The Military Doctrine of the Russian Federation*], No. 146, Moscow, February 5, 2010.

———, *Strategiya natsionalnoi bezopasnosti Rossiiskoi Federatsii*, [*The National Security Strategy of the Russian Federation*], Moscow, 2021. As of October 19, 2021: http://scrf.gov.ru/security/docs/document133/

Selivanov, V. V., and Yu. D. Il'in, "O Vybore Prioritetov Pri Razrabotke Kineticheskogo Oruzhiya Dlya Resheniya Zadach v Voyennykh Konfliktakh" ["On the Selection of Priorities in the Development of Kinetic Weapons for Executing the Missions in the Military Conflicts"], *Voennaia Mysl'* [*Military Thought*], No. 7, 2017.

Shambaugh, David, *China's Communist Party: Atrophy and Adaptation*, Berkeley, Calif.: University of California Press, 2008.

Shmygin, Andrey, "Pul's Planety. Stavka na 'Blestyashchiye Kameshki i Glazki'" ["Pulse of the Planet. Betting on the 'Brilliant Pebbles and Eyes'"], *Krasnaia Zvezda* [*Red Star*], No. 108, 2001.

Sidorov, Vladimir, "Inostrannyye Voyennyye Novosti" ["Foreign Military News"], *Krasnaia Zvezda* [*Red Star*], No. 57, 2005.

———, "Boyevyye Lazery" ["Combat Lasers"], *Krasnaia Zvezda* [*Red Star*], No. 1, 2007.

———, "Pentagon Rvetsya v Kosmos" ["Pentagon Rushes into Space"], *Krasnaia Zvezda* [*Red Star*], No. 46, 2009.

"Silent Contest," transcript, Chinascope, March 5, 2014. As of November 11, 2021: http://chinascope.org/archives/6447

Snidal, Duncan, "The Game Theory of International Politics," *World Politics*, Vol. 38, No. 1, October 1985, pp. 25–57.

"SOI: Bol'shaya Lozh' Reygana" ["SDI: Big Reagan's Deception"], *Krasnaia Zvezda* [*Red Star*], No. 238, 1993.

Sokut, Sergey, "Uroki. Usloviya Prodiktoval Vashington" ["Lessons Learned. The Terms Were Dictated by Washington"], *Nezavisimoe Voennoe Obozrenie* [*Independent Military Review*], June 21, 2002.

State Council Information Office of the People's Republic of China, *White Paper on China's Space Activities in 2016*, December 27, 2016.

Stokes, Mark, Gabriel Alvarado, Emily Weinstein, and Ian Easton, *China's Space and Counterspace Capabilities and Activities*, Washington, D.C.: U.S.-China Economic and Security Review Commission, March 30, 2020.

Stokes, Mark A., and Dean Cheng, *China's Evolving Space Capabilities: Implications for U.S. Interests*, Washington, D.C.: U.S.-China Economic and Security Review Commission, April 26, 2012.

Stone, Alex, and Peter Wood, *China's Military-Civil Fusion Strategy: A View from Chinese Strategists*, Montgomery, Ala.: China Aerospace Studies Institute, 2020.

Su Xiancheng [于小红], Yu Xiaohong [刘震鑫], and Liu Zhenxin [苏宪程], "Analysis of Development of U.S. Space Situation Awareness" ["美国空间态势感知发展分析"], *Journal of the Academy of Equipment Command & Technology* [装备指挥技术学院学报], April 2010, pp. 42–46.

Supryaga, A. V., "O Voynakh XXI Veka" ["On Wars of the 21st Century"], *Voennaia Mysl'* [*Military Thought*], No. 6, 2002.

Swaine, Michael D., "Chinese Views on South Korea's Deployment of THAAD," *China Leadership Monitor*, No. 52, February 2017.

Teng Jianqun [滕建群], "Don't Drive War into Space" ["莫将战车开进太空"], *People's Daily*, November 4, 2006. As of November 11, 2021: http://news.sohu.com/20061104/n246196561.shtml

———, "Why Is China Unhappy with the Deployment of THAAD in the ROK?" China Institute of International Studies, April 1, 2015. As of November 11, 2021: https://www.ciis.org.cn/english/COMMENTARIES/202007/t20200715_2758.html

Tong Zhao, *Trust-Building in the U.S.-Chinese Nuclear Relationship: Impact of Operational-Level Engagement*, dissertation, Atlanta, Ga.: Georgia Institute of Technology, 2014.

Tsygichko, V. N., and A. A. Piontkovskiy, "Dogovor Po PRO: Nastoyashcheye i Budushchey" ["ABM Treaty: Present and Future"], *Voennaia Mysl'* [*Military Thought*], No. 1, 2000.

"Uroki i Vyvody iz Voyny v Irake" ["Lessons and Conclusions from the War in Iraq"], *Voennaia Mysl'* [*Military Thought*], No. 7, 2003.

"U.S. Completes Geosynchronous Space Situational Awareness Program Ground System Upgrade" ["美完成地球同步轨道太空态势感知计划地面系统升级"], *Aerospace Electronic Warfare* [航天电子对抗], Vol. 36, No. 2, 2020.

U.S. Department of Defense, *Quadrennial Defense Review Report*, Washington, D.C., September 30, 2001.

U.S. Department of Defense and Office of the Director of National Intelligence, *National Security Space Strategy: Unclassified Summary*, Washington, D.C., January 2011.

U.S. Department of State Archive, "Strategic Defense Initiative (SDI), 1983," webpage, undated. As of January 21, 2022: https://2001-2009.state.gov/r/pa/ho/time/rd/104253.htm

U.S. Senate, "Testimony on Space Force, Military Space Operations, Policy and Programs," Washington, D.C., U.S. Senate Committee on Armed Services, May 26, 2021.

U.S. Space Command, *Vision for 2020*, Colorado Springs, Colo.: Peterson Air Force Base, 1998.

"U.S. Test-Fires 'MIRACL' at Satellite Reigniting ASAT Weapons Debate," *Arms Control Today*, October 17, 1997.

"V Inostrannykh Armiyakh" ["In the Foreign Armed Forces"], *Krasnaia Zvezda* [*Red Star*], No. 64, 2021.

Valeev, M. G., A. V. Platonov, and V. I. Yaroshevsky, "O Krizisakh vo Vzaimodeystvii Rossii i SSHA v Oblasti Protivoraketnoy Oborony" ["On Crises in the Interaction of Russia and the United States in the Field of Missile Defense"], *Voennaia Mysl'* [*Military Thought*], No. 7, July 2021.

Ventslovskiy, Aleksey, "V Pogone za Zvezdami" ["In Pursuit of the Stars"], *Krasnaia Zvezda* [*Red Star*], No. 211, 2004.

———, "Amerike Nuzhny Den'gi" ["America Needs Money"], *Krasnaia Zvezda* [*Red Star*], No. 21, 2005.

"Voyennaya Platforma Obamy" ["Obama's Military Platform"], *Krasnaia Zvezda* [*Red Star*], No. 207, 2008.

Wang Aijun [王爱娟] and Shi Bin [石斌], "Strategic Debate on the Issues Concerning the Treaty on the Limitation of Anti-Ballistic Missile Systems (ABM) and Missile Defense System Within the U.S. Government (1983–2001)" ["美国政府内部围绕反导条约与导弹防御体系问题的战略论争 (1983–2001)"], *China Military Science* [中国军事科学], January 2015, pp. 123–134.

Wang Chaoqun [汪朝群], "Research on Space Defense" ["太空防御问题研究"], *Aerospace Electronic Warfare* [航天电子对抗], Vol. 25, No. 2, 2009, pp. 19–22.

Wang Shusheng [王树生] and Yang Xuefeng [杨学锋], "'Patriot' Boarded the Ship, and the U.S. Navy's Anti-Missile Capability Was Enhanced" ["'爱国者'上舰，美海军反导能力提升"], *Shipborne Weapons* [舰载武器], September 2008.

Weeden, Brian, "Hearing on China in Space: A Strategic Competition?" testimony presented before the U.S.-China Economic and Security Review Commission, Washington, D.C., April 25, 2019.

———, *Current and Future Trends in Chinese Counterspace Capabilities*, Paris: French Institute of International Relations, Proliferation Papers, No. 62, November 2020.

Wei Chenxi [魏晨曦], "Development Trend of Space Operations from Schriever Wargame" ["从'施里弗'系列演习看未来太空作战的发展"], *Space International* [国际太空], June 2016, pp. 29–36.

Weitz, Richard, "Sino-Russian Cooperation in Outer Space: Taking Off?" *China Brief*, Vol. 20, No. 21, December 6, 2020.

Westwick, Peter J., "'Space-Strike Weapons' and the Soviet Response to SDI," *Diplomatic History*, Vol. 32, No. 5, November 2008, pp. 955–978.

The White House, "U.S. National Space Policy," August 31, 2006. As of October 20, 2021: https://www.hsdl.org/?view&did=466991

———, *National Space Policy of the United States of America*, Washington, D.C., June 28, 2010.

Wolfe, Frank, "Space Force Official Suggests U.S. Developing Offensive Capabilities in Space," *Defense Daily*, October 19, 2020.

Wortzel, Larry M., *The Chinese People's Liberation Army and Space Warfare*, Washington, D.C.: American Enterprise Institute, 2007.

Wu Riqiang, "How China Practices and Thinks About Nuclear Transparency," in Li Bin and Tong Zhao, eds., *Understanding Chinese Nuclear Thinking*, Washington, D.C.: Carnegie Endowment for International Peace, 2016, pp. 219–250.

Xia Yu [夏禹], "Focusing on Future Space Operations: NATO Participates in Shriever 2012 Exercise" ["聚焦未来空间作战:北约参加施里佛 2012 演习"], *Satellite Application* [卫星应用], April 2012, pp. 75–78.

Xiao Tianliang [肖天亮], ed., *Science of Military Strategy* [战略学], Beijing: National Defense University Press [国防大学出版社], 2015.

———, *Science of Military Strategy* [战略学], Beijing: National Defense University Press [国防大学出版社], 2017 revision.

———, *Science of Military Strategy* [战略学], Beijing: National Defense University Press [国防大学出版社], 2020 revision.

Yang Caixia [杨彩霞] and Ai Dun [艾顿], "On the Legality of the Development of ASATs for China" ["论中国发展反卫星武器的合法性"], *Journal of Beijing University of Aeronautics and Astronautics (Social Sciences Edition)* [北京航空航天大学学报（社会科学版）], Vol. 23, No. 2, March 2010, pp. 46–50.

Yastrebov, Yaroslav, "Zametki Obozrevatelya. Vzaimodeystviye Radi Strategicheskoy Stabil'nosti" ["Notes of an Observer. Collaboration for the Strategic Stability"], *Krasnaia Zvezda* [*Red Star*], No. 239, 2001.

"'YEVRPRO': Asimmetrichnyy Otvet" ["European Missile Defense: Asymmetric Response"], *Krasnaia Zvezda* [*Red Star*], No. 222, November 30, 2011.

Yuan Jun [袁俊], "U.S. Space Strategy and Space War Exercise" ["美军太空战略与太空战演习"], *China Aerospace* [中国航天], July 2001, pp. 26–29.

Zhang Jingxu [张景旭], "Progress in Foreign Ground-Based Optoelectronic Detecting System for Space Target Detection" ["国外地基光电系统空间目标探测的进展"], *Chinese Journal of Optics and Applied Optics* [中国光学与应用光学], Vol. 2, No. 1, February 2009, pp. 10–16.

Zhang Ming [张茗], "U.S. National Security Space Strategy Pivot and Its Implications for China" ["美国太空安全战略转向及其对中国的影响"], *Journal of Social Sciences* [社会科学], September 2020, pp. 12–23.

Zhang Qiang, "Why Is America Obsessing over China's Missile Tests?" *S&T Daily* [科技日报], December 7, 2007.

Zhang Xuesong [张雪松], "The United States Builds a Space Force That Has Reached the Point of No Return" ["美国组建天军如箭在弦"], *Space Exploration* [太空探索], September 2018, pp. 56–61.

Zhang Yan [张岩], "India Tests Anti-Satellite Weapon" ["印度成功试验反卫星武器"], *Aerospace China* [中国航天], May 2019, pp. 55–59.

Zhou Lini [周黎妮], Fu Zhongli [傅中力], and Wang Shu [王姝], "Comparison Between Space Deterrence and Nuclear Deterrence" ["太空威慑与核威慑比较研究"], *National Defense Science and Technology* [国防科技], Vol. 36, No. 3, June 2015, pp. 51–54.

Zhu Hui [朱晖], ed., *Strategic Air Force* [战略空军论], Beijing: Blue Sky Press [蓝天出版社], 2009.

Zhu Tingchang [朱听昌] and Liu Jing [刘菁], "Fighting for Command of Space: The Development Process and Influence of the 'High Frontier' Strategy of the United States" ["争夺制天权:美国'高边疆'战略的发展历程及其影响"], *Military Historical Research* [军事历史研究], March 2004.

Zhuang Qubing [庄去病], "Analysis of the U.S. 'Star Wars' Program" "[美国'星球大战'计划剖析"], *International Studies* [国际问题研究], April 1984, pp. 24–36.

"'Zvezdnyye Voyny': Otvetit' Asimmetrichno" ["'Star Wars': Respond Asymmetrically"], *Krasnaia Zvezda* [*Red Star*], No. 164, 2007.

"Zvozdnyye Voyny': Kak SSSR Otvetil Reyganu" ["'Star Wars': How the USSR Responded to Reagan"], *Krasnaia Zvezda* [*Red Star*], No. 169, 2008.